Anthropocene – Humanities and Social Sciences

This series is the first collection to address the subject of education within the Anthropocene. The inspiration behind the series is a dialogue between Earth system science (ESS) and human and social sciences (which, when their object of study is the socio-political implications of our entering the Anthropocene are now officially called the environmental humanities). The series offers perspectives from a variety of disciplines. The raison d'être of this new collection is to draw together high quality studies dealing with the gravity of today's environmental situation whilst also viewing the concept of the Anthropocene from a critical perspective by expanding on work engaged in in human and social sciences over recent years.

The main objectives of this series are to:

- bring together studies from Earth system science and the environmental humanities in the same collection, from an educational angle
- engage with the question of the future and how to prepare for it in a time of uncertainty

This series welcomes topics including, but not limited to:

- pedagogy
- didactics in the Anthropocene
- health education
- education for governance and democracy
- addressing the challenges of contemporary societies and ecological changes
- transforming agricultural production methods
- preparing for the future

The collection would make a valuable contribution to the academic world and research, which is currently seeking new editorial spaces for us to publish scientific works on the implications of entering into a new geological epoch.

Works published in this series include monographs, edited volumes, and handbooks and textbooks.

The series is intended for researcher working on the Anthropocene with all its socio-political implications and also students in different disciplines following courses on environmental issues. It may, however, also be of interest to all social stakeholders interested in environmental questions, who are becoming increasingly numerous including, but not limited to, policy makers, activists, entrepreneurs.

We aim to make a first decision within 1 month of submission.

In case of a positive first decision the work will be provisionally contracted: the final decision about publication will depend upon the result of the anonymous peer review of the complete manuscript. We aim to have the complete work peer-reviewed within 3 months of submission.

Proposals should include:

- A short synopsis of the work or the introduction chapter
- The proposed Table of Contents • CV of the lead author(s)
- List of courses for possible course adoption

The series discourages the submission of manuscripts that are below 75,000 words in length

For more information, please contact the Publishing Editor: Floor.Oosting@springer.com

Thierry Ripoll

Human Psychology and the Destruction of the Planet

Springer

Thierry Ripoll
Department of cognitive psychology
Aix-Marseille Université
Aix en Provence, France

ISSN 2731-6343 ISSN 2731-6351 (electronic)
Anthropocene – Humanities and Social Sciences
ISBN 978-3-032-02079-6 ISBN 978-3-032-02080-2 (eBook)
https://doi.org/10.1007/978-3-032-02080-2

The original submitted manuscript has been translated into English. The translation was done using artificial intelligence. A subsequent revision was performed by the author(s) to further refine the work and to ensure that the translation is appropriate concerning content and scientific correctness. It may, however, read stylistically different from a conventional translation.

This Springer imprint is published by the registered company Springer Nature Switzerland AG
The registered company address is: Gewerbestrasse 11, 6330 Cham, Switzerland

To Jean Rostand, who is undoubtedly the man of letters and scientist who has influenced my view of humanity, nature, and the universe the most.

Foreword

We are heading towards a disaster and we continue to live more or less according to our habits. The privilege of us humans is to know this—and to continue nonetheless on the same path. The big problems will arrive in 10, 20, or 50 years depending on the regions or the GDPs of each. For some, it is already here: fires, shortages, rising waters, population displacement. Year after year, the temperature rises, like that of the broth in which the frog cooks, which at first "finds it pleasant".

This is what makes our time unique! Never has the human being been so aware of his possible extinction, and never has he been so responsible, through his direct action on an entire planet, for what is happening to him.

What do we do with all this? Not much, it must be said. Thierry Ripoll mercilessly points out our compromises and inconsistencies. He depicts a world sick with its obsession with growth, its inequalities, the competition that governs relations between individuals and states, and which pushes us to produce more and more to stay "in the race". Until the final big suffocator, from which only a miracle could save us. You have to read this book to know what it is about. It is a vast, precise, multidimensional panorama of the trap in which *Homo sapiens* has got himself into.

We both come to roughly the same conclusion. We do not spare the planet and living beings without limiting ourselves. Nature dies from man, and if we want it to live, man will have to make himself smaller. Pages of great lucidity show us what is inevitable and what we do not want to see: to preserve the future, you, me, and all of us would have to make great sacrifices. That we accept to significantly reduce our standard of living at our individual level.

Are other solutions possible? Ripoll, as a scientist, is wary of the "all scientific". According to him, relying on technological innovations to save our destiny is an intellectual imposture. For how can we ignore that it is indeed this race for ever more technology that has largely allowed the disastrous situation in which we find ourselves? Science is not responsible, but the immoderate use that humans make of it to satisfy their fundamental urges for pleasure, comfort, and social prestige. As long as the desire for "always more" that drives humanity has not been tamed, no technique will save it.

So, can we imagine a life with less? Fewer trips, less meat, fewer smartphones, less internet connection? In the *Human Bug*, I had mentioned a few ways to find satisfaction in existence despite less outrageous material comfort. Develop mindfulness skills to reconnect with true hedonism—without bulimia for food, travel, or phones. Cultivate an altruism source of cerebral dopamine, and replace material growth with mental or spiritual growth: we will always need to have pleasure; the whole trick is to get it without plundering all our natural resources. That's why the leads sketched by Thierry Ripoll are encouraging. Being a mental being, the human being is equipped to replace the consumption of material goods with play, art, exchange, knowledge as such—and we have known for a few years that these "zero carbon impact" practices have the same effect on the human brain as a hamburger or a large SUV. It's a question of civilisational framework. Hence the certainty displayed in this book: a revolution is needed, because we no longer have time for evolution.

The question remains of the means available to trigger this great shift. Left to themselves, individuals will be too subject to their brain mechanisms of seeking immediate pleasure and ease (the famous striatum lodged in their brains) and to the traps of a globalised society which de facto disqualifies the decreasers by the simple game of competition, to take the initiative of a renunciation of the facilities of the technological society. They will therefore have to be encouraged to do so.

The Individual Ecological Impact Account is a solution proposed by Thierry Ripoll. The remedy is bitter. Such an account would simply allow each citizen to consume and reject waste only to an extent that is compatible with the long-term survival of the planet and its ecosystems. It is therefore a limitation of the desire to consume, to travel, to eat as much meat as one wants, to buy as many cars and phones as one wishes when one has the means. It is, in the end, a deliberate self-limitation of the urges of the most primitive part of our brain that always wants more.

Is it acceptable and legitimate? Many voices will rise up against this infringement of individual freedoms. But in reality, it is placing freedom very low. Sociologists, psychologists, and anthropologists have long known that a human being can only live with others by limiting his own freedom. To socialise is to inhibit oneself. Living together is a series of prohibitions. I must not attack others, I must not steal, I must not run red lights, I must not threaten my fellow men, I must discriminate against them, I must not cross the gates without a ticket, I must not turn on my phone in class, I must not go out beyond the curfew when a virus is circulating, etc. In short, vis-à-vis other humans, *we accept to limit our freedoms*. Simply, we find it intolerable to do so vis-à-vis the planet and species living. Where does this difference in treatment come from? From the fact that we live with a hierarchy of beings, on which we base our actions. In this hierarchy, a river or a species of insect from the Amazon does not have the same rights as a human. It is so because this is in our representation of the world for a very long time.

This representation of the world is obsolete. It is obsolete because it is no longer adapted to the notion of limit that is now imposed on us. The limit is the radically new data that is imposed on humanity. It is an insurmountable obstacle to our delusion of omnipotence. It imposes on us the same transformation as that of the small child who, around the age of one, realises that he cannot do or get everything he wants. This drama of omnipotence is the essential condition for growing up. If he does not pass this stage, the young human never becomes a man.

For humanity, it is simply a matter of becoming human.

de sociologie Sorbonne université
Paris, France

Sebastien Bohler

Competing Interests The author has no competing interests to declare that are relevant to the content of this manuscript.

Introduction

When Awareness of Danger Does Not Lead to Its Avoidance

If we judge by the ability of humans to colonise and exploit almost all terrestrial environments up to the most hostile; if we also judge by the extraordinary increase in their population, there is no doubt that *Homo sapiens* is the very example of an unprecedented evolutionary success. If we add that this incredible trajectory has taken place over an extremely short time in geological terms (3,00,000 years or so if we consider the modern form of *Homo sapiens*) and that, for the most part, it is over the last 10,000 years that humanity has clearly distinguished itself both from its hunter-gatherer past and from our close cousins, the anthropoid apes, then we are indeed confronted with the most incredible, sudden, and fascinating biological and cultural event that our planet has known in its 4.5 billion years of existence.[1] The brevity and the suddenness of the human phenomenon should not make us forget that it is estimated that 50 billion living species have appeared and have disappeared for the most part since the appearance of life on Earth. We therefore constitute an infinitesimal part of the creative power of the living. Not only are we an infinitesimal and contingent expression of the proliferation of life that our universe is capable of generating, we have been occupying this planet as masters for a long period, but above all, we seem to be able to prematurely end our existence while dragging in our wake a considerable number of other species, some of which have been occupying the planet for a few million years. From this point of view, the human species is fascinating, both capable of giving rise to extraordinary civilisations dominating the Earth's ecosystem without sharing and of being the conscious author of absurd destructions that we fear will lead to a profound degradation of the planet.[2] Has evolution given birth to a being as extraordinary and powerful as to be capable of reducing to nothing the diversity of life that it has produced over a few billion years?

[1] Nekola et al. (2017).

[2] Chakrabarty (2018); Glikson and Groves (2016).

What I have just expressed in this introductory paragraph is finally quite banal and perfectly factual. It is likely that a large part of the inhabitants of this planet, especially those who are not totally absorbed by their survival in an environment devastated by war or by poverty, have a clear, sometimes even desperate, awareness of the particular situation in which our species finds itself. Our technological mastery has never been greater, the wealth we produce has never been so disproportionate, the awareness of the limits of our planet has never been so fine, and the information we have about the ecological disaster that is hitting almost all natural environments has never been so precise.[3] Yet, for many of us we have the impression of being caught in a runaway vehicle that is hurtling down the winding road of a mountain at an increasing speed without it being possible either to get out of the vehicle or to slow it down or even to take an escape route. The metaphor is a bit easy I concede. Especially it does not accurately reflect our situation as humans, aware of the danger and the ecological disaster. Indeed, one can suppose that the passengers of such a vehicle are perfectly passive. They have no control over its trajectory. They are inert spectators caught in the horror of a scenario in which they are innocent victims. It is quite different for us today since we are, certainly to varying degrees, actors in this society and involved in one way or another in its functioning as well as in its dysfunction. We participate in the mad race of the vehicle. Some admit it sometimes reluctantly when others manage to cleverly rationalise their actions and behaviours, or even justify them.[4] Others still live this as a devastating cognitive dissonance and develop a deadly guilt. It is a well-established fact: with few exceptions, humans all, to varying degrees, have an awareness of the damage environmental impacts that we produce and the risk that this poses to our species and the entire terrestrial ecosystem.

From a certain point of view, this is the expression of an absolutely unique cognitive and anthropological revolution: a remarkable step that the human species has taken. We have the fortune, or misfortune, to be pioneers in this matter. Indeed, we are the first *Homo sapiens* to collectively realise the finite nature of our planet, the biophysical limits that characterise it, and the impact we have on it. For my own grandparents, born at the beginning of the twentieth century, such a representation of the planet and the role we play on it would have made absolutely no sense: a worldview that would have been totally inaccessible and unintelligible to them. We are therefore generally the first generation to feel some modest or strong guilt, whether it's about taking a plane for a pleasant holiday, living two in a two hundred square metre house, spending the day with our smartphone activated, installing air conditioning, or building a swimming pool... in short, consuming too much in view of the planet's limits. All these behaviours once perceived as pleasant—often ridiculous displays of social success—are now tainted with a feeling of guilt that everyone tries to manage in a more or less honest and effective way. Let's not forget that only a few centuries ago, we still ignored the exact size of the Earth and the very

[3] Barnosky et al. (2012); Steffen et al. (2015).

[4] Gifford (2011).

existence of its five continents. It then seemed so huge compared to the technology of the time that the most perceptive of humans could not for a single second imagine that it was possible one day to destroy it or exploit it to the hilt and ultimately to alter its functioning at the risk of upsetting systemic balances on which we depend. There will indeed be a before and after the twentieth century: the century when humans radically changed their relationship with the Earth. The century when they understood that it is no longer possible to gorge on its prodigality and that the environmental dimension must be integrated into our individual as well as collective projects. Who would have imagined just 50 years ago that tens of thousands of young people would demonstrate, not to earn more, not for more social justice, not for the end of a war, or not to defend any personal interests, but to defend the integrity of a planet in danger? Who would have imagined that the Pope himself would have encouraged them to continue demonstrating? This would have appeared as an unlikely science fiction scenario. Who would have imagined that the concepts of ecology, environmental crisis, and climate change would constitute a significant part of the programmes of political parties, including the most conservative or clearly productivist left-wing parties? No one, I'm pretty sure. I still have a bitter memory of the amused and condescending look that most intellectuals had on the candidacy of René Dumont in the 1978 French elections. We do not really measure the extent and depth of this major anthropological revolution which is only equalled, in my opinion, by the Neolithic revolution. The awareness of the finiteness and fragility of our world will soon have colonised all minds, and we can only rejoice even if such awareness, like most awakenings, is extremely painful.

However, if the awareness is indeed there and if it is manifested loudly, it must be admitted that it is not producing the actions likely to mitigate the environmental disaster even slightly. The facts are not contestable. With a few exceptions—the restoration of the ozone layer for example—the destruction of biodiversity, the destruction of fragile natural environments, and the production of greenhouse gases continue at a frantic pace, without any real respite and regardless of the declarations of good intention or the occasional actions taken here and there by institutions and governments. In short, humanity is at this painful, but necessary, stage that all people suffering from addiction experience. We are for many now aware of the drama that is unfolding, we grasp more or less well the causes, and we know more or less clearly where this is likely to lead us, but we are absolutely incapable of finding the solutions and adopting the behaviours that could protect us from a potentially tragic future. Of course, it would be convenient, and it is common, to blame the powerful of this world, whether they are large multinationals, or states. Certainly their share of responsibility is undeniable and probably in line with the financial and political powers they have. But it would be a very bad strategy to absolve oneself of one's own responsibility by waving the red flag of the "system": the economic, financial, and political powers. In reality, we all actively participate in the environmental crisis that we denounce; even our favourite pets contribute even more than the most poor of this world to the destruction of the planet. As for our political leaders, many are far less Machiavellian than we imagine. I'm not talking about Trump, Bolsonaro, and other brainless leaders who have not yet crossed the anthropological revolution that I mentioned earlier. I'm talking about these typical politicians of our social democracies, concerned with both economic and ecological imperatives. They have no more control than the average citizen over what is happening, firstly because political power has dramatically decreased in the face of the power of the gigantic companies that now dominate the world, and secondly because they are programmed like you and me and this programme does not allow them to conceive of an ecologically compatible society. The result is these speeches full of contradictions in which the same individual—this is also true of the leaders of ecological parties—can at the same time commit to taking measures to revive a growth gasping for breath and lament the loss of biodiversity. At their level, they are absolutely no different from the thousands of French people who despair about the environmental situation and global warming and who air-condition their homes to better cope with the 2 months of heat, thus directly contributing to accelerate the process they denounce: could human rationality just be a mirage?

In fact, we are perfectly stunned, bereft, disconcerted by this situation. We feel powerless, individually and collectively, because taking into account the environmental crisis actually forces us to think about our society in a way we have never thought before. More than thinking about our society, it is the very conception of our life that is concerned, and we do not change as easily as that from a conception of life that has been shaped over the course of a few million years of evolution leading to this curious troublemaker: *Homo sapiens*. What a fascinating and absurd story: biological evolution has produced a living being capable of making disappear what it took a few billion years to produce. From this point of view, the metaphor of the vehicle rushing at increasing speed down a winding mountain road is not so absurd after all. Of course, we are actors in our society and we are not the passive and inert passengers suggested by this metaphor. But the forces, largely unconscious, that exert on all of us, including political or economic decision-makers, are of such power and such universality that they have the power to alienate us to the point of making us as passive, terrified, and toneless as most humans are in the face of an ecological disaster that is becoming more and more precise every day.

Probably, you have wondered what I had in mind when I evoked in a somewhat sibylline way "a conception of life that has been shaped over the course of a few million years of evolution leading to the appearance *of Homo sapiens*". It's actually quite simple. Our psyche is determined both by the cultural forces that exert on us and by the biological forces that we have inherited from our distant hominid past. Even if culture plays a particularly powerful role for our species, we remain largely biologically constrained. Not admitting it would be to deny our obvious part of animality. To take only a few obvious examples, our propensity to love, our linguistic ability, that of memorising or learning, of collaborating, of guessing the intentions of others, of reasoning, of aggressing, or of compassion... are certainly culturally determined, but they are also biologically so. We do not express our love in the same way in Japan and in France just as we do not express our love today as we did in the Middle Ages. Nevertheless, we can identify universals of love, that is to say very general forms of the feeling of love that have, without a doubt, a biological origin perfectly understandable in the context of an evolutionary approach of the human. If we apply this reasoning to the study more generally of the human psyche and the functioning of our society, then we must admit that both are the product of biological and cultural determinisms largely intertwined and most often unconscious. The strong thesis that I will defend in this book is that our human behaviour, the very one that is the cause of the environmental crisis we are experiencing today, finds its origin in the depths of what is probably human nature, biologically and culturally determined. Let's not forget that our past as hunter-gatherers unfolded over about 300,000 years, or even much more if we take into account not the species but the genus *Homo* from which we come. The Neolithic Revolution, the one that led to sedentarisation, to the emergence of villages, cities, empires, private property, technocracy, major armed conflicts, democracy, philosophy, and then science, is a very late phenomenon in human history. As for the industrial revolution, and the enormous technological power that resulted from it, it dates back two centuries: a fraction of a second in geological time. It is very likely that things have gone a little too fast. Under the veneer of high technology as under the gilded panelling of our political and democratic institutions, live men in suits and ties who function hardly differently from our very distant ancestors, or even from our distant cousins: the non-human primates. In short, to paraphrase the great biologist Edward Wilson, we have created a Star Wars civilisation with emotions and behaviours that are those of the Stone Age. In fact, our behaviours, our desires, and our addiction to consumption are perfectly understandable in this theoretical framework. If we cannot change the software that underlies the functioning of our society, it is because it is both inscribed in our genes and subtly integrated into our culture. It is as difficult for us to stop destroying our planet as it is for a smoker to stop smoking because the motives that lead us to destroy it are as opaque and uncontrolled as those that lead the smoker to pay to suffer in a future that he knows is close.

Should we therefore give in to fatalism? Yes and no.

Yes because we are confronted with a society that is a complex system (in the cybernetic sense) extremely flexible and capable of micro-adjustments allowing it to survive without radical rupture. The goal of any autonomous system, like our society, is generally to self-reproduce, to maintain itself by adapting in the least brutal way possible to internal disturbances (technological revolutions or social insurrections) or external ones (global warming for example). There is a psychological bias called the status quo bias. As its name suggests, the longer a situation, even if harmful, persists, the harder it is for us to put an end to it. The same applies at the societal level. To give a familiar example directly related to the subject of this book, the green economy and the development of non-fossil energy sources are the very examples of these micro-adjustments, undoubtedly useful in the short term, but strictly incapable in the longer term of resolving the aporia we must face: continuing to grow in a finite world.

Yes, also because we are individually the actors of this system and we contribute despite ourselves to its perpetuity. What's more, even when we oppose it, we keep it alive at the same time. While denouncing it, even loudly or violently, we support it with our behaviour. Few are the revolutionaries, advocating a vague and hollow "return to nature", who would really be ready to concretely assume what it would mean to lose all the advantages that we individually derive from maintaining the society we denounce... especially when we are among the privileged of this planet. Like these dogs that show their teeth and bark furiously behind the fence of the property they guard and who walk away with their tails between their legs when they are directly confronted with the intruder, we are quick to virulent criticism, but so incapable of acting in coherence with this criticism in our concrete actions of our daily life.

Yes, finally because, as indicated above, the forces that lead to the maintenance of this system are so numerous, so obscure, so unconscious, so deeply buried in our psyche, that there is something almost inevitable about participating in its maintenance whatever our displayed and conscious will to question it. In fact, I consider that the situation in which our society is and its inability to manage the environmental crisis is "normal". It is normal, not in the sense that we should be satisfied with it and accept it, but in the sense that, given the determinisms of different natures, first biological and psychological but of course then economic, sociological, geostrategic, anthropological, philosophical, religious that constrain us, a situation other than this one would have been surprising. The objective of this book will therefore essentially be to identify these determinisms. This seems essential to me because it seems unlikely that we will be able to free ourselves from them as long as they continue to exert their power over us without us even being aware of it.

And yet no. We must not give in to fatalism. Firstly, because if one is never certain of winning when one decides to fight, one is certain of losing when one gives up. Above all, and more fundamentally, the fact that we are subject to such deep determinisms, so anciently inscribed in our history as *sapiens* both as a biological and cultural entity, does not mean that our path is traced and that it inevitably leads us towards chaos. Humans have an extraordinary adaptive capacity and an unimaginable collective intelligence that have enabled them to effectively face very hostile environments. We must therefore take the risky bet that they will be able to overcome this critical stage of their brief history without the price to pay being exorbitant either for the planet as an ecosystem or for humans as a particularly special host of this ecosystem. Do I have some solutions to propose for this? Probably a few, probably some leads. Certainly no miracle solution, simplistic, caricatured such as the champions of radical utopia claim to hold and which consists in naively thinking that once the system has been overturned, we will finally be on the happy path of justice, peace, and respect for the environment. But the essential is not there. The essential will be to identify the power of the psychological or anthropological obstacles that prevent us from adopting a lifestyle compatible with the preservation of the planet and ourselves. Because I will show in this work, the solution is not technical, far from it. It takes a lot of naivety, *hubris*, pretension, arrogance, and probably incompetence to imagine that the solution will come from our scientific and technological mastery. The solution will necessarily result from a deeply different relationship to oneself and the world.

References

Barnosky, A. D., et al. (2012). Approaching a state shift in earth's biosphere. *Nature, 486*, 52–58.

Chakrabarty, D. (2018). The seventh history and theory lecture anthropocene time. *History and Theory, 57*(1), 5–32.

Gifford, R. (2011). The dragons of inaction psychological barriers that limit climate change mitigation and adaptation. *American Psychological Association, 66*(4), 290–302.

Glikson, A. Y., & Groves, C. (2016). *Climate, fire and human evolution: The deep time dimensions of the anthropocene*. Springer.

Nekola, J. C., Allen, C. D., Brown, H. J., Burger, J. R., Davidson, A. D., Fristoe, T. S., Hamilton, M. J., Hammond, S. T., Kodric-Brown, A., Mercado-Silva, N., & Okie, J. K. (2017). The Malthusian–Darwinian dynamic and the trajectory of civilisation. *Trends in Ecology & Evolution, 28*(3), 128–130.

Steffen, W., et al. (2015). Planetary boundaries: Guiding human development on a changing planet. *Science, 347*(6223), 1259855.

Contents

Chapter 1
For a Brief Overview

Abstract For numerous cognitive reasons, it is extremely difficult to grasp the severity of the environmental destruction affecting our planet's ecosystem. In this chapter, I first outline some characteristics of the human cognitive system that make a realistic perception of the planet's ecological situation impossible. Subsequently, I analyze the most notorious and dangerous degradations in the most objective way possible, in light of what is commonly referred to as the planet's limits.

1.1 The Difficulty of Perceiving Environmental Destruction

Our cognitive system is the result of a few million years of evolution and the cognitive skills that are ours today differ almost not at all from those of our ancestors from 100,000 years ago. Among the skills that have certainly contributed to the unchallenged domination of our species over all others, is our ability to plan and anticipate: the temporal horizon of a human far exceeds the temporal horizon of the most brilliant chimpanzee.[1] Not only can we conceive of a relatively distant future, including one that is after our death, but we can also remember our own past and imagine that of our ancestors. Humans have the ability to live as much in the present time as in the past or future. Without this ability, *Homo ergaster*, a very distant ancestor from over a million years ago, would never have been able to plan the development of the first sharp biface allowing him to hunt or butcher prey that were not even present at the time of the development of his weapon.

Nevertheless, our brain is not particularly well equipped to handle long periods of time and we encounter great difficulties in concretely and effectively

[1] Even if ethology research has highlighted astonishing mnemonic and anticipatory abilities in anthropoid monkeys.

T. Ripoll, *Human Psychology and the Destruction of the Planet*, Anthropocene – Humanities and Social Sciences, https://doi.org/10.1007/978-3-032-02080-2_1

representing temporalities that exceed those we face in our daily lives.[2] For millions of years, like all animals, most of our goals were short-term goals and even though we are much more efficient than other animal species in this respect, it is very difficult for us today to project ourselves concretely even at the twilight of our lives. We can think about the future in an abstract way, but so abstract, so theoretical and conceptual, that it seems very distant to us. Too distant in reality to encourage us to change our current behaviour and influence the decisions we make.[3] This, we will see, is one of the reasons why the most devastating addictions are so difficult to overcome. When a smoker smokes, when an alcoholic drinks, when an obese person gorges themselves, the effect of the immediately pleasant action weighs heavily compared to the distant and terrifying negative consequences that inevitably result from their addiction. Moreover, none of these negative consequences are perceptible at the time. The catastrophic long-term effect has little impact on the decisions we make because it is in a future that remains too abstract to be properly considered. The same is true of our consumption, which is the primary cause of the planet's destruction and constitutes a form of universal addiction from which we must free ourselves. Each act of consumption, even the most ordinary, is likely to provoke a small immediate pleasure. This small pleasure is cognitively immediately accessible. The immediacy of the satisfaction it provides crushes the distant consequences that such consumption has on our environment with its concrete force.

The cognitive limitations of humans mean that we are not equipped to properly apprehend either too large quantities (can we represent, for example, a distance of one light year, the mass of the sun: 1.9891×10^{30} kg or even the amount of CO_2 emitted per year: 38 billion tonnes?) or extreme temporalities (can we represent the 15 billion years of existence of the known universe, the 4.5 billion years of existence of the earth or even the 300,000 years of existence of our species?). Our brain has not evolved to allow this simply because our survival in our natural environment *of Homo sapiens* does not require the mastery of such a capacity. As a result, we perceive very poorly the enormous changes we are making to our environment over a long period of time. If old photographs of natural sites are so captivating and informative, it is precisely because they compensate for our inability to remember the visual characteristics of places that are familiar to us. What a surprise, what astonishment when we can simply see the place where we live as it was 20 years ago! What a shock when the superimposition of old and new photos helps us to become aware of the impact of global warming on our glaciers or on the ice cap! Our brain integrates the new characteristics of the most familiar landscapes at such a prodigious speed that their often progressive and small-step destruction ends up completely escaping us. Obviously, this type of cognitive limitation contributes to the non-recognition of the destruction of our planet and makes it extremely difficult to correctly appreciate the monstrous degradations we have been inflicting on it for about 50,000 years at least. This is why I believe that the only way to truly measure the environmental disaster of which we are the sole actors is precisely to emphasise the temporal dimension of this degradation, which I will try to do in this chapter.

---

[2] Myong-Il and Shinsuke (2016); Ainslie (2001).

[3] Weber (2006).

1.2 A Demographic Explosion Unique in the Animal World

Let us first recall that from a strictly demographic point of view, it is estimated that for a large part of the existence of the genus *Homo,* its population was of the order of 10,000 individuals. To give a comparable order of magnitude, the current population of the two subspecies of gorillas in great danger of extinction, yet confined to a restricted area of equatorial Africa, is about 60,000 individuals! We have long been on the brink of extinction! 50,000 years ago, when our species probably directly or indirectly eradicated the other species of the genus *Homo* with which it cohabited and at the time when it colonised most of the planet, we were about 1.5 million individuals,[4] the population of a small Chinese city spread over five continents. An extraterrestrial naturalist exploring planet Earth at that time would have had great difficulty crossing paths with the rare animal that we were then. The human population then slowly increased and temporarily stabilised until the Neolithic revolution, about 10,000 years ago, a crucial time when humans settled down and gradually developed agriculture and livestock. From then on, the population will grow exponentially: a phenomenon of exceptional intensity without equivalent in the rest of the animal world. 100 million humans around −2000, 200 million at the end of antiquity, 500 million around the fifteenth century and the billion mark reached only at the beginning of the fourteenth century. This is already extraordinary, but is nothing compared to the increase of these last two centuries, even worse of these last 50 years, that is to say a fraction of a second in the light of the history of our species: 1.6 billion in 1900, 2.5 billion in 1950, 6.1 billion in 2000 and 7.8 billion today. These orders of magnitude are absolutely fascinating and worrying. In the last fifty years, our population has grown more than during the 300,000 years of existence of our species hence the term "great acceleration", the post-war period during which the population increased by 3 billion in just 50 years. Even though the phenomenon of demographic transition has led to a slowdown in growth, the latter remains very significant and the most rigorous anticipations are based on a population of about 11 billion in 2100, an expected increase of 40%. Of course, human density has evolved in proportion[5]: 1 inhabitant for 3300 ha in the Palaeolithic, 1 for 100 ha in the Neolithic, 1 for 10 ha in the Iron Age and 1 for 2 ha today. To this should be added the billions of domestic animals (livestock or pets) which constitute a kind of extension of the human species and which play a considerable role in the imbalance of the Earth system. The factors behind this population explosion are multiple: climatic, cultural, biological, economic and especially technological. In Fig. 1.1, the sudden demographic growth is related to the technological revolutions that have marked our history. This clearly illustrates the recentness of the population explosion even though the point at the origin of this graph is already very late (the beginning of the Neolithic around 10,000 years ago). As mentioned above, this dizzying demographic growth does not jump out at us because we are cognitively

[4] For all these demographic data, see Biraben (2003).

[5] Jouventin (2016).

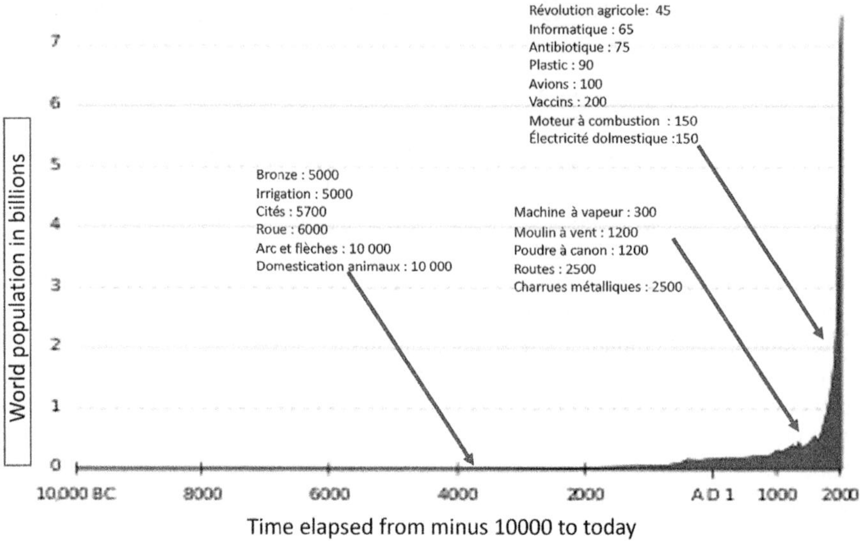

Fig. 1.1 Increase in the world population between −10,000 and 2000 and date of different technological innovations (Graph taken from Nekola et al. (2017))

unable to grasp its brutality: our cognitive horizon is too narrow for that and nothing in the personal experience of our brief lives allows us to have a concrete experience of it. We suffer from a bias that prevents us from grasping exponential trajectories. They are so counter-intuitive that we cannot become aware of the terrifying scale of the phenomena in progress. Often, awareness comes very late and occurs at a time when it is no longer possible to control the phenomenon. That said, for any ecologist, it is clear that no natural system in balance is capable of absorbing without major destabilisation the proliferation of a species, let alone a species as productive and energy-consuming as ours.

1.3 Energy When You Hold Us

Naturally, given the gigantic ecological footprint of a single contemporary human, our impact on the planet has reached hallucinating proportions. This impact being roughly the product of the number of inhabitants by their consumption, the curve of this impact is even more extreme than that of demographic growth. Again, it is probably very useful to clearly identify the evolution of a human's energy consumption throughout our history. To fully grasp the problem, we must consider that the Earth system is a relatively closed system that receives energy from the sun and expels some of it into space, the differential can be stored or consumed. We must imagine the Earth as a gigantic battery and, like any battery, it can store energy, produce it and lose it. At first, the planet was biologically dead because no living organism could either consume or produce energy. It took 1 billion years for life to appear in the form of basic unicellular beings now capable of producing and consuming energy. They are the ones who started to transform solar energy and returned it to the planet in different ways. The global charge of the Earth battery then began to increase. The phenomenon generally exploded with the large-scale development of photosynthesis. The energy balance of the earth then stabilised except for exceptional upheavals (great volcanic activity, climatic fluctuations, collision with giant asteroids). Overall, for 4.5 billion years, the charge of the Earth battery has greatly increased. This energy has been partly stored in the form of fossil energy (gas, coal, oil…) and in the form of living biomass, mainly forests and all terrestrial plants as well as the animals that depend on them. This balance was perfectly preserved even when humans appeared on earth. The energy he exploited then, in the form of withdrawals from gathering, hunting, burning wood necessary for cooking, heating or protection had no impact on the global energy balance. It was only with the appearance of livestock and agriculture, about 10,000 years ago, that the balance was gradually broken. Our Paleolithic ancestors (before 10,000 BC) generally consumed 2000 Kcal per day. This consumption increased significantly in the Neolithic and has literally skyrocketed over the last two centuries, particularly over the last 50 years. It is now considered[6] to be 24 times greater than in our distant past, with huge variations within the global population as it has hardly progressed for most inhabitants of poor countries while it has increased by a factor of 100 for those in rich countries.[7] This results in a major energy imbalance unique in the history of the planet, a fundamental imbalance that will ultimately lead to the simple energetic and physico-chemical impossibility of ensuring the survival of our species.

It is possible that the analysis in terms of energy flow and storage is too abstract to immediately grasp its importance, but we must never forget that the human being is a material being consuming energy without which its survival on Earth is simply impossible. We can discuss the orders of magnitude and quantitative estimates always suffer from some inaccuracies, but from a simple and fundamental physical point of view, the energy trajectory we are following is simply a highway to the precipice because we consume infinitely more than what the regenerative capacity of the planet allows.[8] We are reproducing on a planetary level and for all of humanity what flourishing civilizations have regionally done before definitively disappearing.[9] In reality, just in the last century, we have consumed far more energy than during the entirety of our existence on Earth. Currently, especially in terms of fossil energy, we are burning what the Earth has accumulated over hundreds of millions of years. This is nothing more or less than a radical holdup, an inconsequential and perfectly irrational gluttony that starkly illustrates the inability to anticipate the

[6] Schramski et al. (2015).

[7] The energy consumed by a wealthy Westerner is 1000 times greater than that consumed by a modest inhabitant of the Sahel according to Satterthwaite (2009).

[8] Steffen et al. (2015); Brown et al. (2011).

[9] Diamond (2005); Tainter (1988).

future, especially the fate of future generations. Such is the characteristic of this addiction to consumption: making our descendants pay for the recklessness of our present behaviour.

1.4 Some More Concrete Forms of Energy Imbalances

The particular case of phytomass (the mass of all plants) is quite worrying, especially because the amount of phytomass conditions the food supply of humans and the billions of farm animals they depend on. It constitutes the most sustainable and renewable way to recharge the Earth's energy and it allows the development of animal biomass on which the general balance of our ecosystem depends. Different calculation methods all converge towards the same result. Although we have modified about 85% of the planet's surface for our food needs—if we do not count the surface of ice-covered lands—by destroying most of the flora and fauna that lived there, we are approaching the breaking point, that is, the point where the total arable surfaces of the Earth will no longer allow the survival of our species. To give an idea of the scale of this destruction, the stock of phytomass has been reduced by 45% over the last 2000 years and during the twentieth century alone, we destroyed 17% of the phytomass that existed at the beginning of the century.[10] In gross values, the phytomass at the beginning of our era was about 1000 GT[11] of carbon, it was reduced to 660 GT in 1900 and to 550 GT in 2000. This destruction is measured not only in mass and surface destroyed but also in productivity. The latter has collapsed and is often artificially maintained by overdoses of all kinds of amendments that further accelerate soil degradation, in addition to their overall impact on chemical and atmospheric pollution. Although the data are indisputable and governments have been widely alerted, this massive destruction has accelerated even more in the last ten years, raising fears of exceeding a threshold beyond which repair possibilities become nil. This situation is all the more masked and difficult to grasp as waste remains massive and many agricultural productions are destroyed to maintain a good level of profitability.

The situation is obviously little different at the level of the zoomass—the global mass of living beings not including plants—especially when compared to the anthropomass (the mass of humanity). The latter simply quadrupled in the twentieth century while at the same time the zoomass was reduced by 50%, now representing a very small part of the total biomass. Moreover, the biomass of farm animals has largely exceeded the biomass of wild animals. Thus, there are about 1.4 billion cows in the world and 150 billion farm animals are killed each year for our food. While the biomass of domestic animals has increased by 350% over the last millennium, that of wild mammals has collapsed, so much so that the former represents 25 times

---

[10] Smil (2011).

[11] GT is the symbol for gigatonne, or 10^9 tonnes.

the biomass of all wild mammals. Overall, humans has appropriated the bulk of phytomass and biomass and has played a crucial role in the transformation of the Earth. This has resulted in a terrifying degradation of natural ecosystems and the emergence of ecosystems that appear natural to us but are in fact heavily impacted by human activity. It is estimated today that three quarters of the planet have been directly modified by human activity and that humans have appropriated 90% of the overall productivity of the planet.

1.5 Consequences in Terms of Biodiversity

We can lament the loss of biodiversity for two reasons of a completely different nature, referring to two quite opposing philosophical conceptions.

Either we lament it in a very anthropocentric way because it is ecologically perfectly demonstrated that the greater the biodiversity of an ecosystem, the more resilient it is and therefore capable of adapting and resisting internal or external disturbances. Preserving biodiversity is a way of preserving ourselves. This is, for example, the low biodiversity of intensive agriculture that makes it so fragile in the face of pests or cryptogamic diseases. Preserving the biodiversity of the planet therefore constitutes a form of buffer capable of mitigating environmental shocks affecting the Earth system, whether these are of anthropogenic origin or not. Contrary to the particular sensitivity that humans have towards living things, especially the fact that they are particularly moved by the disappearance of animals as emblematic as certain species of large mammals, the disappearance of much more discreet species located at the base of the food web is much more dramatic and worrying from this point of view. The disappearance or rarity of plankton, as well as the disappearance of bees and other pollinating insects or that of earthworms could immediately endanger the very existence of humans on Earth. In terms of ecology, the most worrying disturbances are often the least immediately perceptible, our survival depending more on barely visible insects than on species as emblematic as polar bears, gorillas or whales. From this point of view, we must preserve biodiversity because it conditions the survival of humans on Earth.[12]

But we can also lament the loss of biodiversity because it is nothing more or less than a form of holocaust[13] or ethically problematic ecocide that has this major characteristic of being irreversible: a species that has disappeared is gone forever. It is considered that it takes 5–10 billion years for the level of biodiversity to be restored after a phase of mass extinction, which is about 20,000 times the time we will have taken to reduce this biodiversity to nothing if we continue at the current rate for a few more centuries. The objective of preserving biodiversity results more from ethical considerations than utilitarian ones. It reveals a form of biophilia that has greatly

[12] Burch-Brow and Alfred Archer (2017).

[13] Wilson (1999); Crist (2007).

increased in recent years and has led to questioning the privileged place that humans have on Earth.

Whatever the case, there is no longer any doubt that we are experiencing the sixth mass extinction that the planet has known, the only one, however, that is of human origin. The biologist E.O. Wilson, not without humour, has proposed to call this new era, the Eremozoic era, literally, the era of emptiness or loneliness since most of the living species of the current environment have or will disappear. The extinction of species is a perfectly natural phenomenon that results either from environmental changes, or from the evolution of a species that adapts to environmental changes and gradually disappears by evolving towards another phenotype. Simply, the normal rate of disappearance has absolutely nothing to do with the current rate of disappearance. Based on palaeontological investigations, this rate of disappearance for fauna has been estimated at 0.5 extinctions per million species per year.[14] Currently, one million species are threatened with extinction[15] and 26,000 species disappear each year. This rate is expected to accelerate dramatically with the impact of global warming and this is all the more so as the migratory routes that animal species naturally take when environmental conditions are no longer favourable can no longer be taken due to the fragmentation of natural spaces. More than 30% of mammals, amphibians, reptiles and insects are threatened with extinction by the end of the century, a millisecond in geological time.

Two phenomena that have been widely popularised and publicised should not constitute the tree that hides the forest. These are, on the one hand, the relative "good health" found in emblematic species and the increase in protected areas such as national parks and reserves. In the first case, this includes, for example, the increase in the population of the common rorqual or the humpback whale in the Atlantic, the griffon vulture and the bearded vulture in Europe, the return of the wolf and the increase in the number of ungulates in France. But for a few expanding species that have been subject to specific protection measures, there are hundreds of others, often less spectacular, that are disappearing to the point that some birds once commons have become rare. In agricultural areas, a third of bird populations have thus disappeared. As for national parks, they constitute islands of protection whose interest cannot be disputed but which have become vast open-air zoos, sometimes delimited by immense fences that illustrate a terrible fragmentation of habitat. We should not be deceived by the overabundance of certain species (the herring gull in Provence, the wild boar or Florida turtles in our ponds) which owe their demographic dynamism only to the ecological imbalances for which humans are solely responsible. Overall, on a planetary scale, as in our own country, the faunal heritage is shrinking at such a speed that it is dramatically given to any naturalist to experience it concretely even in the short span of time that represents a human life.

---

[14] Mace et al. (2005).

[15] 2019 Global Assessment Report on Biodiversity and Ecosystem Services published by the Intergovernmental Science-Policy Platform on Biodiversity and Ecosystem Services

1.6 The Limits of the Planet

Chemists, biologists and ecologists have converged towards a quantitative approach to the planet in terms of limits,[16] limits that cannot be exceeded without seriously endangering its balance and therefore the survival of most of its inhabitants including humans of course. These limits concern ten fundamental properties among which we find: the acidity level of the oceans, the climate, the level of the ozone layer in the atmosphere, the phosphorus cycle, the nitrogen cycle, biodiversity… According to these researchers, each of these properties of the system interacts with the others as is necessarily the case with any complex self-regulated system. As a result, when a limit is crossed for a given property, it impacts each of the other properties hence the possibility of non-linear effects in the sense that the effects potentiate and can have massive consequences far greater than the sum of their effects. Let's just take one example to illustrate this phenomenon: that of the nitrogen and phosphorus cycle. Nitrogen is a gas present in the atmosphere which is then transformed by bacteria that make it assimilable by plants. Nitrogen is therefore a basic element of most fertilisers as it is a necessary element for plant growth. Phosphorus exists in solid form and is directly extracted from the earth's crust in the form of phosphate. Like nitrogen, phosphorus is necessary for plant growth. The cycle of these two chemical elements has been totally disrupted by the intensive use we make of them in agriculture. 120 million tonnes of nitrogen present in the atmosphere are ultimately used each year for the manufacture of industrial fertilisers which are then discharged into natural environments, rivers and oceans while the bearable limit for the planet is estimated at 35 million tonnes. 20 million tonnes of phosphate are extracted from the earth's crust for the same reasons, of which nine million tonnes flow into the oceans even though the limit would be one million tonnes.[17] This is not insignificant. It should be recalled that in the past, palaeontological records have shown that mass extinction of life in the oceans corresponded to excessive levels of phosphorus. Current models show that an excess of 20% of the normal amount of phosphorus is sufficient to cause mass extinction in the oceans. In short, for these fundamental constituents alone, the limit has clearly been exceeded. Taken individually, the crossing of the limit in terms of sustainability by the Earth ecosystem has a considerable impact in the immediate and very dangerous in the longer term, but taken together, their effects are simply catastrophic. Thus, the disruption of the nitrogen and phosphorus cycle combined and their discharge into the oceans or rivers leads to the progressive anoxia of the oceans, which is already total in areas close to rivers or in relatively closed seas (Baltic Sea, Gulf of Mexico, bottom of the Sea of Cortez…). This anoxia alters these ecosystems to the point of making all life impossible. The effect is deleterious as it interacts with the effects of global warming. Indeed, global warming linked to the overproduction of CO_2 leads

[16] Rockström (2009); Steffen et al. (2018).

[17] The details of the calculation of these values and their analysis can be found in the work of Rockström's team (2009).

to the acidification of the oceans which, in turn, contributes to the decrease in the amount of plankton. However, plankton plays an essential role in the production of oxygen and in the absorption of CO_2. Chemical pollution, that linked to phosphorus and nitrogen—we are not discussing here the impact of many other pollutants such as pesticides for example—by weakening the natural balance of the oceans increases its vulnerability to the danger induced by acidification caused by the increase of CO_2 in the atmosphere. There is a set of very complex feedback loops, often difficult to model precisely, but which obviously converge towards a disruption of the balance of the Earth system of which we are obviously beginning to perceive the consequences. It is clear that the continuation of our activity at the same rate as currently leads us straight to major ecological upheavals too rapid and too massive for any solution to protect us from an ecological disaster of which we will of course be among the victims.[18]

1.7 Global Warming as Seen by Researchers and as Seen by the Public

The subject is so well-trodden that it might seem unnecessary to address it here. Nevertheless, I will do so because it is the object of impostures and conspiracy theories contributing to the chronic disinformation that leads the general public to doubt everything and, by doubting everything, to delay the moment when concrete actions must be taken to end the infernal cycle of destruction in which we are engaged. It happens that I have long been interested in the study of climate and that I am interested as a cognitive psychologist in the cognitive processes that lead to the adoption of unfounded beliefs.[19] So I have carefully studied not only the scientific work in the field, but also the research conducted by psychologists on the establishment of unfounded beliefs about the climate. These are very numerous and, having explored almost all the publications accessible to the general public and the websites dedicated to this subject, I must admit that such a situation is not surprising. It is serious and worrying because if some unfounded beliefs have little impact on the functioning of our societies, those concerning global warming have a major impact.

Surveys conducted among the public reveal remarkable scepticism not only about the reality of global warming but also and above all about its anthropogenic origin. It must be recognised that purely ideological or conspiratorial sites, often produced by so-called scientists whose incompetence is matched only by their arrogance, are so well made that it is normal for a reasonably sceptical citizen to question the reality of the phenomenon. When, moreover, politicians, either ignorant in the field or subtly opportunistic, display without any restraint their own climate scepticism—this is of course the case of the last American Presidents except Obama

[18] Steffen et al. (2018).

[19] Ripoll (2020).

but also of the majority of nationalist and populist leaders—it is a boulevard that opens up to climate scepticism. In reality, however, the consensus among scientists is around 97%.[20] It has indeed continued to grow as scientific investigations have been carried out in this field for over 50 years now. This figure of 97% results from the analysis of 12,000 articles published in the leading journals in the field and is beyond any possible dispute. The 3% of researchers who are still sceptical are either particularly cautious researchers who believe that the very high probability of anthropogenic warming is not yet sufficient to adopt a categorical position, or researchers, most often from the private sector, who have been corrupted by governments in order to instil doubt in the general public. This strategy has been clearly adopted by the American governments and methodically organised and financed by the major oil groups, notably Exon.[21] It has indeed worked wonders since, particularly in the USA, only 13% of the population is aware that the consensus among researchers is over 90%. For all Western countries, citizens believe that 67% of researchers agree on the anthropogenic origin of global warming, a figure far removed from the actual consensus of 97%. In short, the majority of the planet's citizens are not aware that the consensus is total among scientists: global warming is undeniable and its anthropogenic origin leaves no room for doubt.[22]

1.8 Global Warming: A Chance and a Disaster Foretold

The title of this section may surprise. How could global warming be a chance for humanity and the planet? It is a chance because by directly threatening the global economy and our way of life, it constitutes the spur that allows the acceleration of the awareness of the need for radical changes at least in terms of energy consumption and probably in terms of the overall functioning of our societies. Indeed, if global warming did not directly threaten us, it is highly likely that we would continue with business as usual. Neither the destruction of natural spaces, nor the loss of biodiversity, nor even the extravagant level of pollution in large metropolises and its impact on the health of billions of individuals would lead to questioning the current model. The other "virtue" of global warming lies in its global character. Even though global warming does not manifest itself with the same intensity in different regions of the globe and even though its impact is not of the same nature in the Sahel zone and in Siberia, it remains a planetary phenomenon that, in the long term, should massively impact all nations. Let us recall in this respect that according to

---

[20] Cook et al. (2013); Cook and Jacobs (2014).

[21] Oil companies have used the same techniques as those used by tobacco manufacturers in the 60s by circulating an online petition (the global warming petition project) of which 99.9% of the signatories were fake. Scientists or even worse, fictitious people.

[22] Durand (2018).

the already old report by Nicholas Stern,[23] a loss of 5–20% of global GDP and therefore of about 10,000 billion dollars could be caused by global warming if nothing is done to bend the CO_2 emission curve. According to the former President of the World Bank, the economic impact of this phenomenon by 2050 should be worse than that of the 1929 crisis and the two world wars combined. Our blindness is so powerful that only a direct threat to our economic model and our dependence on the sacrosanct growth could provoke a real consideration of the environmental dimension: global warming has undeniably played this role. From this point of view, it is a chance in the same way that suspicion or the announcement of cancer can help a smoker to free themselves from their addiction.

Let's now return to global warming itself. It is, of course, an extremely complex phenomenon that is caused by such a quantity of factors that its modelling remains very difficult. This explains why climate predictions are always tainted with a margin of error that cannot challenge the reality of the phenomenon. Everyone knows that the Earth's climate has undergone numerous upheavals, both caused by variations in the level of solar radiation and natural production of greenhouse gases. What is certain, however, is that despite the existence of geological, biological and astronomical factors that naturally impact the climate, it is indeed the increasing emission of human-made greenhouse gases that is the main cause of a sudden warming of the climate that has had no equivalent in the history of the planet. It should also be noted that the rate of CO_2 in the atmosphere, which has naturally varied over time, has been the major cause of the hot or cold periods of our distant past. For example, the increase in CO_2 caused the hot episode of the Cretaceous and the transition from the Paleocene to the Eocene, the temperature then increasing by 6 degrees and causing a rise in sea level of 15 metres. What is absolutely terrifying is the amplitude and speed of climate change. The emission of CO_2 is currently increasing at a rate of 20 ppm per decade, a rate 100 times higher than during the last 800,000 years. The average temperature is rising at a rate that even the most alarmist forecasters had not included to the point that even the average person can perceive a change that normally unfolds over geological times infinitely too long to be identified in the short span of a human life.

This is of course reflected in the global phenomenon of melting glaciers, but also in the displacement of fauna and flora. A study conducted in Europe and the USA has shown that a very large number of species have moved north at an average rate of 17 km per decade, while movements during previous phases of natural climate change reached 40 km per century. In the Mediterranean, the surface temperature is experiencing an exceptional increase, leading new species of subtropical or tropical origin to colonise this sea with its unique ecosystem. The speed of the phenomenon is therefore without comparison with previous phases of warming or cooling and

[23] The Stern report has of course been criticised and the assessments on the economic impact of global warming are orders of magnitude. It should simply be kept in mind that even if we admit the very approximate nature of its anticipations, the economic impact of the upcoming climate shock will be absolutely gigantic.

has already led to the disappearance of a large number of species that do not have the possibility of evolving adaptively on such a reduced time scale.

The gravity of the phenomenon also lies in its non-linear and threshold nature. Non-linearity means that the increase in temperature is not in a simple and constant relationship with the increase in the level of CO_2. Beyond a certain threshold, a runaway of the phenomenon will occur, leading to an even more brutal and considerable rise in global temperature. Thus, we know that an increase of 3 degrees will cause the total melting of the Greenland ice cap. An increase of 4 degrees should destroy 80% of the Amazon rainforest and cause the melting of the Siberian permafrost. All these upheavals, for different and perfectly well identified reasons, will further accentuate the more direct effect of greenhouse gases to the point that the rise in temperature should disrupt the whole of biological life and the functioning of our societies within only a few decades. Let us remember that major changes in climate have been the primary cause of species extinction and the collapse of great civilisations.[24] On the other hand, we must take into account the enormous inertia of the phenomenon in the sense that the disturbance of the climate manifests itself rather slowly at first due to the very size of the planet, but its effects accelerate and continue long after the process has been initiated. A quarter of the CO_2 currently emitted will still affect the climate after a thousand years and a tenth after a hundred thousand years, so that a return to a normal situation could take nearly a hundred thousand years.[25] This is indeed a recurring problem with our anticipations. Our models of the amount of CO_2 in the atmosphere rarely go beyond 2100. However, according to the most rigorous studies, by 2300, the level of CO_2 in the atmosphere should reach the incredible value of 1400 ppm, which would lead to an increase in temperature of about ten degrees, making our survival on Earth almost impossible.[26] 60% of the calories consumed by humans result from three cereals: wheat, maize and rice.[27] However, a temperature increase of just 1 degree would lead to a massive reduction in productivity that could no longer be compensated by the expansion of cultivated areas. A recent study conducted for Australia showed that a 2-degree increase would cause a 50% drop in cereal production.[28]

Climatologists have been making predictions for over 50 years now and these predictions have all been greatly exceeded by the reality of the phenomenon. Its acceleration has surprised even the most pessimistic forecasters. Today, the leading experts in the field agree on the existence of several scenarios that take into account the uncertainties related to the complexity of the phenomenon and the evolution of human-induced greenhouse gas production. Whatever these scenarios, including the most optimistic ones, all show that even if drastic CO_2 reduction measures were

[24] Diamond (2005).

[25] Summerhayes and Zalasiewicz (2018).

[26] Bala et al. (2005); Kasting (1998).

[27] Battisti and Naylor (2009).

[28] Asseng et al. (2011).

taken today, the increase in temperature by the end of the century would be such that it could cause both a collapse of human societies and major ecological upheavals drastically and definitively reducing a large part of the current biodiversity.

1.9 The Indisputable and the Disputable

Unless one is in serious denial, or perhaps also fails to recognise or accurately measure the severity of the environmental situation and its consequences for mankind, it must be admitted that this extraordinary ecosystem that is our planet, isolated in a vast and desperately empty space, is on the brink of an unprecedented crisis. "Being on the brink of an unprecedented crisis" is probably an understatement or a worn-out and somewhat unrealistic slogan because obviously, we are already in the crisis although for most of us, we are barely aware of the drama unfolding before our eyes for the simple and stupid reason that our individual lives are not yet directly and visibly impacted. We are no different in this respect from smokers who, a few days before the announcement of a cancer that has already begun its destructive and fatal work, continue to slowly commit suicide, sometimes even mocking the risk they are running. The human cognitive system needs the spectacular to react and immediate spectacular. Without it, its inertia is such that inaction will always be its favourite response. The covid 19 crisis has disrupted human activity across the planet in a matter of weeks and governments have made decisions that no scenario, however crazy, could have incorporated. Yet, "only" 100,000 people have died to date from the virus in France, mostly elderly people or those with comorbidity. Yet 60,000 French people die each year in relation to air pollution alone and infinitely more if we took into account other environmental disturbances. But these deaths are transparent for the simple and good reason that they are spread out over time. Yes, therefore, the ecological disaster we are experiencing is indisputable, but for a host of strictly psychological reasons, we are neither able to measure its magnitude nor to react rationally even though we have become aware of what is at stake. Indeed, for most scientists in the field, we have entered a new geological era: the Anthropocene. The Anthropocene follows the Holocene and is distinguished from all other geological eras by the fact that it is human who is, for the first time in the history of the planet, the origin of a geological era. Our environmental impact has been so immense that we have managed to dangerously alter the general balance of the planet on most of the physical and biotic parameters that characterise it. A future palaeontologist examining the terrestrial environment would easily identify the occurrence of a catastrophic event, a cataclysmic discontinuity whose beginnings can be identified at the start of the Neolithic with an incredible acceleration from the twentieth century. Human is now a major geophysical force that rivals natural geophysical forces to the point of becoming a direct threat to the global environment.[29]

[29] Steffen et al. (2018).

Evolution seems to have produced a species with cognitive abilities such that its expansion and domination now seem on the verge of causing an unprecedented environmental crisis, perhaps the equivalent of the great Jurassic extinction.

As I indicated at the beginning of this chapter, the difficulty we encounter in grasping the magnitude of the disaster is due to the fact that we are incapable of grasping its rapidity: everything seems to be going very slowly whereas in terms of geological time, which is the time of our planet, everything is evolving at an incredibly fast pace. Our perception of time, our perception of the major changes for which we are responsible is not compatible with a correct appreciation of the ongoing phenomenon. Geological time, the time of the evolution of a system as vast as our planet is a time that escapes us. There is therefore a terrible paradox here. Our daily lives seem so little impacted by the enormity of the ongoing phenomenon that we underestimate its power and speed. Yet it is this speed that is problematic because it makes it impossible for living beings to adapt through the classic evolutionary processes that require long times to be implemented, just as it makes it very unlikely that our species will adapt to the global upheavals for which we are responsible. The other dimension which completely eludes us concerns the inertia of the Earth system, this inertia being inherent to its size and the complexity of the dynamic processes that contribute to its apparent stability. The state of the Earth today is already the result of events that have unfolded over tens of thousands of years. This is particularly the case with the loss of biodiversity as is the case with global warming, both of which have been caused by human presence for over 40,000 years. In other words, our behaviour today determines some of the biotic characteristics that will be those of the Earth in a few hundred or thousand years: a future infinitely too distant to be reasonably taken into account by a brain of *Sapiens,* more concerned, like any modestly equipped animal for this, with its present than with a future it struggles to grasp.

All of this constitutes the undeniable part of this story. But what is the disputable part? There are two from my point of view.

The first is of little importance. It is simply about approximations related to the projections that scientists develop from models that are certainly imperfect but generally correct. This type of uncertainty is actually perfectly negligible because, over the long term that should concern us, errors in projection, even significant ones, would change absolutely nothing about the problems we are going to have to overcome. The economist Tim Jackson[30] reminds us in this regard that if growth remained in the years to come at the level it was over the last 50 years, the average level of consumption and predation on the planet in 2100 would be 100 times higher than it was in 1950. Even if for various reasons, the pace of growth will not maintain at this level, our productive, consumer and destructive capacity will far exceed the absorption capacity of the planet. At the current rate, or even at a much lower rate, several hundred planets would be needed to absorb the impact of human activity in just a few centuries. We can therefore dispute the detail of the analysis made by scientists;

[30] Jackson (2009).

we cannot dispute the fact that the current trajectory is simply absurd,[31] unrealistic, utopian and fundamentally selfish in view of the world that future generations will inherit.

The second is more subtle. It is the hypothesis that it is possible to maintain the level of growth necessary for the proper functioning of the global economy (about 3%) without this leading to the destruction of the ecosystem on which we are dependent. This would require "simply" decoupling our productive activity and consumption from the environmental damage currently associated with it: a green and highly technological growth that would have the incredible advantage of not changing anything about our current operation while preserving the environmental integrity of the planet. It's a grand, magnificent and yet totally utopian bet that I will analyse later.

1.10 What's All This for?

Why are we destroying the planet and with it our chances of long-term survival? The question may seem laughable but deserves consideration. Fundamentally, because humans aspire to improve their standard of living. But why do we aspire to this? This is the heart of this book and the deep reason for our mad race towards a catastrophe that is rapidly approaching. The reason for this incessant movement towards a legitimate search for improvement in our living conditions can be explained in multiple ways. But it is likely that the reason that everyone immediately has in mind is: because we are all seeking happiness and the most immediate way to achieve it is precisely to improve one's own living conditions. We must therefore now study the determinants of happiness.

References

Ainslie, G. (2001). *Breakdown of will*. Cambridge University Press.

Asseng, S., Foster, I., & Turner, N. (2011). The impact of temperature variability on wheat yields. *Global Change Biology, 17*, 997. https://doi.org/10.1111/j.1365-2486.2010.02262.x

Bala, G., Caldeira, K., Mirin, A., Wickett, M., & Delire, C. (2005). Multicentury changes in global climate and carbon cycle: Results from a coupled climate and carbon model. *Journal of Climate, 18*, 4531–4544.

Battisti, D., & Naylor, R. (2009). Historical warnings of future food insecurity with unprecedented seasonal heat. *Science, 323*, 240–244.

Biraben, J.-N. (2003). The rising numbers of humankind. *Population & Societies, 394*, 1–4.

Brown, J. H., Burnside, W. R., Davidson, A. D., Delong, J. P., Dunn, W. C., Hamilton, M. J., Mercado-Silva, N., Nekola, J. C., Okie, J. G., Woodruff, W. H., & Zuo, W. (2011). Energetic limits to economic growth. *Bio Science, 61*, 19–26.

[31] Tainter (1988).

Burch-Brow, J., & Alfred Archer, A. (2017). In defence of biodiversity. *Biology and Philosophy, 32*, 969–997.

Cook, J., & Jacobs, P. (2014). Scientists are from Mars, laypeople are from Venus: An evidence-based rationale for communicating the consensus on climate. *Reports of the National Center for Science Education, 34*, 3.1–3.10.

Cook, J., Nuccitelli, D., Green, S. A., Richardson, M., Winkler, B., Painting, R., Way, R., Jacobs, P., & Skuce, A. (2013). Quantifying the consensus on anthropogenic global warming in the scientific literature. *Environmental Research Letters, 8*(2), 024024.

Crist, E. (2007). Beyond the climate crisis: A critique of climate change discourse. *Telos, 141*, 29–55.

Diamond, J. (2005). *Collapse: How societies choose to fail or succeed*. Viking.

Durand, F. (2018). Global warming: Colossal stakes for new societal projects. *Parcours*. Cahiers du GREP Midi-Pyrénées, GREP MP. hal-02004842.

Jackson, T. (2009). *Prosperity without growth: Economics for a finite planet*. Earthscan.

Jouventin, P. (2016). *Man, this failed animal*. Free & Solidarity.

Kasting, J. (1998). The carbon cycle, climate, and the long-term effects of fossil fuel burning. *Consequences, 4*, 15–27.

Mace, et al. (2005). Biodiversity, Ch. 4. In H. Hassan, R. Scholes, & N. Ash (Eds.), *Ecosystems and human wellbeing: Current state and trends* (pp. 79–115). Island Press.

Myong-Il, K., & Shinsuke, I. (2016). Time discounting, present biases, and health-related behaviours: Evidence from Japan. *Economics and Human Biology, 2*(1), 22–136.

Nekola, J. C., Allen, C. D., Brown, H. J., Burger, J. R., Davidson, A. D., Fristoe, T. S., Hamilton, M. J., Hammond, S. T., Kodric-Brown, A., Mercado-Silva, N., & Okie, J. K. (2017). The Malthusian–Darwinian dynamic and the trajectory of civilisation. *Trends in Ecology & Evolution, 28*(3), 128–130.

Ripoll, T. (2020). *Pourquoi croit-on ? Psychologie des croyances*. Auxerre, Sciences Humaines.

Rockström, J. (2009). A safe operating space for humanity. *Nature, 461*, 472–475.

Satterthwaite, D. (2009). The implications of population growth and urbanisation for climate change. *Environment & Urbanisation, 21*, 545–567.

Schramski, J., Gattie, D., & Brown, J. (2015). Human domination of the biosphere: Rapid discharge of the earth-space battery foretells the future of humankind. *Proceedings of the National Academy of Sciences, 112*(31), 9511–9517.

Smil, V. (2011). Harvesting the biosphere: The human impact. *Population and Development Review, 37*(4), 613–636.

Steffen, W., et al. (2015). Planetary boundaries: Guiding human development on a changing planet. *Science, 347*(6223), 1259855.

Steffen, W., et al. (2018). Trajectories of the earth system in the anthropocene. *PNAS, 115*(33), 8252–8259.

Summerhayes, C. P., & Zalasiewicz, J. (2018). Global warming and the anthropocene. *Geology Today, 34*(5), 194–200.

Tainter, J. A. (1988). *The collapse of complex societies*. Cambridge University Press.

Weber, E. U. (2006). Experience-based and description-based perceptions of long-term risk. *Climatic Change, 77*, 103–120.

Wilson, E. O. (1999). *The diversity of life*. Norton.

Chapter 2
A Magnificent and Irresistible Decoy: The Search for Happiness

Abstract To truly understand the deep-seated reasons for our inability to rationally, reasonably, and effectively address the environmental crisis, we must be aware of the psychological determinants that shape our behaviors. At the heart of this issue is our universal quest for happiness and our misguided conceptions about it. Here, I present empirical studies by psychologists and economists on happiness, particularly analyzing the complexity of the relationship between consumption levels and happiness.

At the end of the confinement period linked to the covid 19 pandemic, I met one of my friends who had taken advantage of the time he had to get into permaculture. He informed me with the utmost seriousness that confinement had allowed him to distance himself from society and, in the end, he had decided to leave it, ready to abandon his house for a hut, and clearly decided to consume no more than the bare minimum for his food and his austere future housing. At the time of our meeting, I had with me a photographic trap that I use regularly to study the presence of nocturnal mammals. When he noticed it and I explained its function, he asked me how much I had paid for it and promised to buy two or three to discreetly observe the animals frequenting his plot of land. I pointed out the obvious contradiction between his vow of frugality and his project to purchase high-tech equipment of more than questionable utility. He was surprised by my remark and, with a twist that we are all familiar with, explained to me that this was an unimportant exception that did not challenge his overall approach. In my innermost self, I was amused by an obvious contradiction to which I myself hardly escape. And as always in such circumstances, I wondered about the increase in happiness that this new acquisition was really going to bring him. There seems to be the crux of the problem: what does our happiness depend on and does it vary in proportion to the consumer goods of which we are so fond? After all, we are in the process of destroying the planet because our

T. Ripoll, *Human Psychology and the Destruction of the Planet*, Anthropocene – Humanities and Social Sciences, https://doi.org/10.1007/978-3-032-02080-2_2

propensity to consume requires us to exploit its resources more and more each day. This is undoubtedly regrettable, but if this destruction at least has the merit of making us happier then this destruction presents a minimal and indisputable form of rationality echoing the classical economic conception according to which well-being and happiness are a direct consequence of the accumulation of wealth, a conception that Diderot himself had already formulated in a most condensed manner: "The whole economy of human society is based on a simple, fundamental and general principle: I want to be happy…". If it turned out that this happiness-wealth relationship was not real then it would be a lose-lose situation. In reality, it would simply be dramatically absurd: destroying the planet to increase our level of wealth without it making us happy would be the height of absurdity. The study of the determinants of happiness, beyond being a major research theme in psychology, is thus politically and ecologically essential if we admit that the ultimate object of all politics is to enable the greatest number to achieve an optimal level of happiness. Far from distancing us from the political and ecological question, this chapter will constitute one of the cornerstones of the developments that will follow.

Research on happiness has seen a very significant development over the past 50 years. They have been conducted mainly by psychologists, neuroscientists, anthropologists and economists. The starting point of these investigations is a famous article dating from 1974 published by an economist. This is Richard Easterlin, the originator of the no less famous Easterlin paradox.[1] Easterlin conducted a very large empirical study covering many nations.[2] The individuals in these surveys had to locate their level of happiness[3] on a so-called Likert scale comprising 10 levels, 0 meaning very unhappy, 10 totally happy for example. Happiness being perfectly subjective, the only way to measure it objectively is to ask people to evaluate their overall level of satisfaction, happiness or well-being.[4] The first result of this survey is that there is a weak, but real relationship between the GDP[5] of nations and the level of happiness of its inhabitants: citizens of rich countries consider themselves happier on average than citizens of poor countries. That said, happiness or well-being are very complex states, difficult to grasp and resulting from multiple factors. This explains that, there is indeed a statistical link (a correlation) between happiness and GDP, but this relationship is not systematic and quite weak in view of the considerable differences in wealth between countries such as Norway or Senegal for example. Thus, Costa Ricans are generally happier than Americans even though the GDP per capita in the USA is four times higher than the GDP per capita in Costa Rica. Overall, the happiest inhabitants of the planet are Europeans and among these,

[1] Easterlin (1974).

[2] In a 2015 article, Easterlin extended his study to 37 countries.

[3] "Happiness" is here the translation of "happiness". The terms "welfare" or "well-being" are frequently used in this literature. They refer to a global estimate of the level of satisfaction one has in their life. To be brief, we talk about happiness, but rigour would have required us to talk about "subjective feeling of happiness".

[4] I will use the terms "happiness" and "well-being" interchangeably.

[5] GDP, acronym for Gross Domestic Product, is an economic indicator that measures the wealth production produced by all economic agents of a country.

the palm goes to the inhabitants of Scandinavian countries, far ahead of the USA. The unhappiest inhabitants are those of African countries. Moreover, it is also observed that this relationship between wealth and happiness is very marked at the individual level. Within a nation, the rich are happier than the poor. To summarise the result of Easterlin's work, at time T when people are asked, there is a weak, but undeniable, link between level of wealth and happiness, which seems to contradict the popular saying that "money can't buy happiness". On the other hand, and this is where the paradox lies, over a given period, there is no relationship between the increase in wealth and happiness. Concretely, over a period of 10 years for example, the increase in a nation's GDP has no significant impact on the level of happiness of its inhabitants. Easterlin[6] in 2010 specifically analysed the case of China, Korea and Chile. These three countries have seen their income per capita double, in less than 10 years for China, in 13 years for Korea and in 18 years for Chile. On the other hand, the level of happiness of their inhabitants has slightly declined. In the same way, the level of happiness of American citizens has not increased in 70 years while the income per capita has increased by more than 60%.[7] Overall, therefore, the richer one is, the more likely one is to be happy, but becoming richer does not significantly contribute to being happier. We now need to explain this paradox.

2.1 The Reasons for the Easterlin Paradox

First, let's start with a brutal evidence with consequences probably more formidable than it appears. There can be no linear relationship between increase in wealth and happiness, simply because while it is possible to infinitely increase an objective datum (wealth expressed in income, GDP or whatever you want), this is impossible for a psychological property such as happiness. What could it possibly mean, and would it even be possible, to increase one's happiness level by 100, 200 or 300%? Let's illustrate this with a thought experiment in which we would assume that per capita income or even GDP increases by 2% for 1000 years, an order of magnitude that is quite reasonable economically speaking. After 1000 years, the GDP would be 400 million times larger than it was at the start, with the associated disastrous environmental consequences of such a level of production and consumption. It is of course unimaginable that individual happiness could increase in the same proportions unless one were to live, at every moment, a form of ecstatic orgasm that would undoubtedly be exhausting. It follows that if there were to be a positive correlation between GDP and happiness, it should tend towards 0 over time. We must therefore conclude that over the long term, there is absolutely no relationship between levels of wealth and happiness. And yet, I am convinced that the individuals of this imaginary land of plenty would continue, as frantically as we do, to do everything to become even richer even if they knew that this would have no impact on their level of happiness. Are we driven by desires unrelated to what is likely to make us happy?

---

[6] Easterlin et al. (2010).

[7] Easterlin (2017).

This is to be feared and what I describe here is one of the possible explanations for the Easterlin paradox. The empirical results obtained by many authors have largely confirmed this theoretical speculation. They observed that, beyond a certain threshold, any further increase in wealth had no further effect on happiness.[8] However, for all rich countries, this threshold has long been exceeded, which is why individuals in rich countries are happier than those in poor countries, but no longer progress up the happiness scale, or even regress, even as their wealth continues to increase.

The second explanation for the Easterlin paradox lies in the fact that the impact of wealth is relative, not absolute. To put it in a slightly caricatured way, it's not being rich that makes you happy, but being richer than your neighbour. In other words, an increase in wealth could lead to more happiness provided that others become less wealthy than oneself.[9] As Gérald Bronner reminds us,[10] "In all areas, comparison seems to hurt the eye of the one for whom it is unfavourable". This seems quite crude and ridiculous. We will see in the next chapter that this is an unavoidable element of human nature, a sad and ridiculous vestige of our distant and still so present past as *Sapiens*. This results in seemingly irrational choices. In multiple experiments conducted by psychologists,[11] subjects were asked to choose between two types of situation that we present here in a simplified way. They had to choose, for example, between working in company A and being paid $35,000 when their colleagues earned $38,000, or working in company B and being paid $33,000 when their colleagues earned $30,000. The vast majority chose company B, preferring to opt for the lower salary as long as others would be paid even less than themselves. In a field study[12] this time, such a phenomenon was confirmed by the fact that at equal salary level and identical socio-professional status, people living in poor neighbourhoods were significantly happier than those living in rich neighbourhoods. Something similar was found from the study of the satisfaction felt by high-level athletes who participated in the Olympic Games.[13] Quite astonishingly, silver medallists are less satisfied with their performance than athletes further from the podium. When their psychological state is analysed, it is identified that their disappointment comes from the comparison they cannot help but make with the competitor who won gold. The mere fact of imagining that they could have won gold is enough to generate a frustration that ends up greatly attenuating the happiness that should have resulted from an already exceptional performance. "Don't look in your neighbour's plate" is an injunction that almost all parents give to their children, even though they are probably the first not to respect it in their daily lives. This is probably an injunction they sense as salutary, especially as they are quite incapable of adhering to it. In a less politically correct register, the obsession with "the biggest" that so many men covet, far from being grossly vulgar, is the perfect expres-

[8] Kerschner (2010); Kahneman and Deaton (2010).

[9] Easterlin et al. (2010); Layard et al. (2010).

[10] Bronner (2021).

[11] Griffin and Tversky (1992); Solnick and Hemenway (1998); Carlsson (2007).

[12] Luttmer (2005).

[13] Medvec et al. (1995).

sion of an intraspecific competition that makes comparison between individuals the alpha and omega of a race lost in advance towards happiness. Humans are eminently social animals and comparison with others remains a major determinant of our sense of success, accomplishment and happiness. The problem is that if the engine of our actions in order to be happier is largely induced by the In comparison to others, the race for more wealth (or more social status) that results could well resemble a zero-sum game, that is to say a game where the sum of the gains and losses of all individuals equals 0. This is exactly what the Red Queen expressed in her response to Alice who was wondering about the need to run even faster to go elsewhere: "here, you see, one has to run as fast as one can just to stay in the same place. If you want to go somewhere else, you have to run at least twice as fast as that". If this analysis is valid, and it certainly is given the number of empirical studies that have validated it, the game is not quite zero-sum. The big loser is first our environment which pays the price for our consumption frenzy and the stupid race we all impose on ourselves to enrich ourselves more than others. The other big losers are obviously those who are on the wrong side: the losers, the poor, the left behind... The most absurd thing about this phenomenon is what the Anglo-Saxons have called the "positional treadmill". This metaphor means that, since everyone tries to take advantage of the other and the others do roughly the same thing, the relative position of each one hardly changes. What a tragic race seems to be imposed on humans! For some to see their happiness increase, others must succeed less well than them, but as the obsession with comparison remains constant, even those who have gained a few places will consider themselves too far from the leading pack that stretches to infinity. The distress of the losers is not even compensated by the happiness of the winners. This phenomenon is remarkably illustrated by the constraints of sports competition, with the difference, however, that given the small number of competitors at the highest level, there is at least one winner, or even a few winners who will temporarily benefit from their success. Since by definition, athletes live in and from competition, they are very often condemned to doping. In fields such as cycling where performance is almost exclusively dependent on the physiological capabilities of the riders, it is no longer possible to win a race without being doped. The practice of doping is even observed among amateurs who are not subject to the requirement of professional success. Obviously, both spectators and runners would like doping to disappear. But since the race is positional and no one can be sure that his competitor is not doping, almost everyone agrees to ingest drugs with deleterious effects. No one therefore derives any benefit from it since, everyone proceeding in the same way, or almost, everyone ends up in theory on an equal footing: this is a very concrete expression of what a zero-sum game is. From this senseless and archaic race, institutionalised and sometimes sophisticated expression of this search for success whose theoretical objective is to make us happier, all participants are ultimately losers. The winner has won by cheating and this is likely to haunt him in his conscience, if not in public life. The others will have just seriously endangered their health without gaining any real advantages. Those who have refused doping are out of the game. They become the downgraded of the cycling society, honest victims of an absurd system. As for the spectators, they tire of a corrupt system, a tragic allegory of the functioning of our own societies. And yet the system continues without ethical speeches and coercive or control measures really managing to change it: the power of psychological forces, and not just economic ones as is often believed, are probably much greater than we imagine.

The race for happiness would therefore also be a race for inequality and probably for injustice. The irony is that the very existence of inequality could be a necessary condition for the hope of being happy since being happy seems to be able to occur only from the moment one considers oneself rich among the poor. In a certain way, the radical egalitarianism advocated by communist countries was a response to the tragic and absurd nature of such a dynamic. But by imposing strict egalitarianism, this ideology deprived men of what constitutes the engine of life for most of them: the natural propensity to want to succeed better than others in the hope of gaining a few points on the happiness scale. The hallucinating amount of medals and decorations that the citizens of communist countries wore during festivities or ceremonies was just a modest symbolic substitute for the luxury items that Westerners like to wear to distinguish themselves socially and display their social and professional success even though it does not make them progress on the happiness scale. Medals or Louis Vuitton bags, the essential thing is to have them when others do not.

The last explanation of the Easterlin paradox is based on a fundamental psychophysical phenomenon: habituation. If it is 40 degrees outside and you enter a room where it is 30 degrees, you will perceive it as cool and very pleasant. But if soon a few minutes have passed, you will estimate that its temperature is finally not very pleasant and still too high. Conversely, if it is 30 degrees and you plunge into water at 22 degrees, the latter will seize you and seem too cool. But after a few minutes, it will seem quite bearable. In the same way, if you put an object on your hand, you will immediately perceive its weight. But after a few minutes, you will have no sensory awareness of it. By getting used to its presence, the object will have disappeared from our consciousness. Extended to higher level phenomena, this adaptive phenomenon has dire consequences. When a pleasant event occurs, it has an immediate effect on the level of satisfaction, well-being or happiness, but quite quickly, time producing its effect, one finds oneself at the initial level of happiness. The Anglo-Saxons have called this phenomenon "the hedonic treadmill", literally the hedonic treadmill which I would translate as the concept of hedonic adaptation. This phenomenon occurs in both directions. An unfortunate event can cause great sadness, depression, intense suffering. Yet, a few months are usually enough to regain the level of happiness that preceded the unfortunate event. Thus, studies conducted[14] on people who were suddenly and permanently disabled revealed—quadriplegic people or even people suffering from locked-in syndrome—that most of them regained a level of happiness equivalent to that which preceded the onset of the disability a few months later. However, when able-bodied people are asked about their reactions and general psychological state if they were to become seriously disabled, all respond that they would be severely depressed, a significant number even specifying that life would not even be worth living if the disability was too

[14] Brickman et al. (1978).

severe. Obviously, they are mistaken. In the same way, but conversely, shortly after their marriage, spouses experience a quite intense state of grace. But, even in the case of a successful marriage, their overall level of happiness returns to its initial level after an average of 3 years, often well before.[15] The case of lottery winners is even more remarkable. The intense feeling of happiness following the receipt of a huge win is followed by a very rapid return to the initial state, and even, in most cases, a catastrophic drop in the initial level of happiness. In 2015, a field survey was conducted among Kenyan villagers.[16] Researchers granted additional income to certain families. Interviewed a few days later, these families rated their level of happiness more positively than families who had not benefited from this additional income. They clearly indicated that they were more satisfied with their lives and less stressed, and this was even manifested by a decrease in cortisol levels (cortisol is a hormone produced in the presence of stress). However, a year later, all these positive effects had definitely disappeared even though their level of wealth was higher than it would have been without the gift that was given to them. Only one negative effect remained: the subjective happiness reported by the villagers who had not benefited from this gift had significantly decreased.

This phenomenon is as basic as it is relentless and everyone has been able to experience it very concretely in very ordinary situations. When men had the opportunity to travel by horse, they were probably fascinated by the speed at which they could travel. When cars replaced horses, the latter seemed terribly slow. When cars reached 130 kilometres per hour, driving at 90 kilometres per hour was hardly more bearable and once you have travelled from Marseille to Paris by TGV or plane, the length of a fast car journey no longer seems tolerable. The same is true of the satisfaction provided by most of our new material acquisitions: larger house, more comfortable car, more powerful computer, more efficient camera... Personal happiness is thus confronted with a limit that moves away as soon as we approach it or believe we have approached it, propelling us into an endless race leading to the exhaustion of the most credulous and naive. This is manifested in an almost comical way, if it were not dramatic in some respects, for people wishing to acquire real estate. They all estimate on average that they would need just 30% more than their current budget—regardless of course of the budget they have—to acquire the "dream" home. The same phenomenon is observed when it comes to estimating the level of salary that would make life really comfortable and pleasant.

These three phenomena that explain the Eastrerlin paradox are very powerful. Even when we become aware of their existence, it is difficult to get rid of them because they draw on the deepest human psychology.[17] They ultimately combine to propel us into this futile race, which consists not only in trying to distinguish ourselves from others by a surplus of social success, but also the one that inevitably

---

[15] Lucas et al. (2004).

[16] Haushofer and Shapiro (2015).

[17] Benjamin et al. (2012); Diener et al. (2013, 2018); Hsee and Hastie (2006).

traps us by encouraging us to accumulate more and more wealth to accumulate more and more consumer goods. They are at the foundation of consumerist and liberal ideology, notably hedonic habituation and the relative nature of happiness, which is why we can politically reject consumerism and liberalism and submit to it in our daily behaviour like these people dependent on a drug they condemn and yet consume cheerfully. Indeed, surveys conducted among European and American citizens all go in the same direction. Although they admit that their consumption is excessive, this does not translate into concrete actions in their own lives. On the contrary, the most educated and richest people, those who are most aware of the impact of excessive consumption on global warming for example, consume in a completely compulsive way.[18]

2.2 The Determinants of Happiness

The title of this section only makes sense if we presuppose that happiness is obtained in the same way for everyone, for all times and all cultures. Of course, this is not the case and the forms of happiness have varied greatly historically, just as they still differ quite widely today in Asia and the West. However, studies conducted in psychology allow us to identify universals, particularly in terms of the factors that determine the personal level of happiness.

Research conducted in recent years agrees that happiness is the product of the combined effect of three factors[19] which have been identified and quantified from statistical analyses allowing the level of happiness to be related to multiple objective data relating to the individuals studied. These are 1/ the base level, 2/ the circumstances and 3/ voluntary actions to increase one's level of well-being.

The base level would be genetically determined. We do not all have the same predispositions to happiness. It is estimated that the part of variance (i.e. the variability of happiness) explained by this factor is at least 50%. In other words, among all the factors that determine the level of happiness, genetic heritage would account for at least half. This has, among other things, been confirmed by the study of homozygous twins (true twins).[20] Each person would have a variable base level, which can temporarily evolve according to circumstances, but which quickly returns to its initial level, the latter being conceived as a powerful attractor from which it is difficult to escape. The existence of a genetically determined base level is perfectly consistent with the concept of hedonic adaptation that we have developed above to explain the Easterlin paradox. The occurrence of positive or negative events impacts our perception of happiness, but this impact is much less durable and solid than we

[18] Bowerman (2014).

[19] See Caunt et al. (2013) for a synthesis.

[20] Lucas and Donnellan (2007).

generally imagine because, as soon as their effect has passed, we return to the base level that is unique to us.

The second factor (circumstances) contributes in a much more modest way, of the order of 10–20%, to our feeling of happiness. By circumstances, we mean the stable characteristics of our situation: our health, our marital status, our social status, our age, our sex, our place of living, our height, our salary, our level of wealth... Among the characteristics of the "circumstances" factor, being married and having a job play the most important role. One might be surprised at such a weak effect of this factor, but this is largely explained by the phenomena of hedonic adaptation and systematic comparison of our situations, which we have seen are at the origin of the Easterlin paradox. There is a very particular characteristic of this factor which partly explains the importance we attach to it despite its low capacity to impact our feeling of happiness. This is a bias, highlighted in cognitive psychology and behavioural economics, called loss aversion. This bias leads us to be much more sensitive to losses than to gains.[21] In other words, a loss of an amount X will have a much greater effect on our psyche than a gain of the same amount. This is quite terrible psychologically because it implies that a strong increase in one's standard of living is needed to barely and temporarily increase one's level of happiness, while a small loss of standard of living has a major and lasting impact, which is why salary cuts are much more psychologically devastating than increases are beneficial. This is how economists have shown, from a panel including a large number of countries, that the level of happiness is five times more sensitive to the contraction of income than to their increase.[22] This also explains why periods of recession have a major impact on the level of well-being while periods of strong growth have a negligible impact. In a certain way, loss aversion could justify the fact that we rarely abandon the race for more wealth, more social status and more consumption. Because if winning a few places in this imposed race does not change much in our lives, losing them can have a deleterious effect.

The third factor (voluntary actions) accounts for about 40%, which means that there is room for manoeuvre, allowing one to voluntarily modify one's level of well-being. This is a broad category in which we can integrate actions as diverse as: having physical activity, maintaining intellectual activity, developing creative activities, getting involved in a political movement, developing social relationships, pursuing goals in line with one's own values... These are all the actions we can undertake, which do not have a direct impact on what we have called "circumstances", but which play a major role in our feeling of happiness. Among the multitude of voluntary actions grouped in this category, two play a determining role: those that determine the quality of our social relationships and those that contribute to giving life meaning. Sébastien Bohler[23] has brilliantly shown how the loss of meaning can profoundly affect an individual's psychic state. All other things being equal,

---

[21] Kanheman and Tverski (1979).

[22] De Neve et al. (2015).

[23] Bohler (2021).

Moreover, people who engage in an associative or political movement, those who actively participate in a religious or charitable movement… feel happier than others. The fact of identifying a meaning, even if it is fictitious, in our actions, massively contributes to raising our level of happiness. In the same way, the more people have strong and satisfying social ties, the more they have positive romantic relationships, the more they have solid and stable friendships, the more they cooperate with others, the more they are happy. The quality of the relationships we have with other humans, our social capital, is therefore a major determinant of happiness, which is not very surprising given the eminently social dimension of our species.

2.3 For GDP to Increase, the Trap Must Close

If the determinants of happiness are indeed those we have just identified then we may well be collectively trapped by the sirens of a happiness which, like a seductive—or seducer—perverse, makes us glimpse ecstasy and pretends to let itself be caught to slip between our fingers just when we think we can grasp it. This relentless mechanism is exploited at will by our liberal societies, capable of fuelling an infinite desire that can never be fully satisfied, simply because the very object of desire is, by nature, incapable of quenching this insatiable thirst for happiness. On this subject, Gérald Bronner in "The Cognitive Apocalypse" quotes this remarkable phrase from Charles Kettering, influential member of General Motors: "The key to prosperity is the creation of organised dissatisfaction". Indeed, we all put considerable energy, sometimes to exhaustion, into changing what we have called above circumstances while these play a modest role in our feeling of well-being or happiness. A large part of our life is motivated by the prospect of increasing our standard of living, our social status and our salary in order to give us the possibility to consume even more. The trap then strengthens in multiple ways to the point that only a few marginal people sometimes escape from it. Indeed, since the mad race that carries us towards the goal of more wealth or social status is a race that humans mercilessly deliver, it results not only in constant and brutal competition, a growing level of inequality and the perception of others as a potential threat to their own success. These elements are also major determinants of happiness, but negative ones. Indeed, the more a society is built on a high level of competition between individuals, the less happy the citizens are overall, probably because extreme competition creates a climate of uncertainty and anxiety that is not compatible with any form of psychological well-being. Moreover, even when the level of wealth increases significantly, this has no impact on the level of happiness if wage and wealth inequality also increases, so that there is a negative correlation between the Gini coefficient[24]

[24]The Gini coefficient is a statistical measure used to evaluate the level of inequality within a population.

and the level of happiness.[25] This is not very surprising either because the rich then perceive the poor as a threat and the poor suffer from their humiliating condition. Perhaps this also explains why the level of happiness is higher in states that have higher levels of taxation allowing to reduce the level of inequality, as is the case in Scandinavian countries.[26] Finally, the compulsive search for increased wealth and consumption of material goods within a society of generalised competition is not conducive to voluntary actions that contribute most to the increase of happiness: the establishment of strong social ties and engagement in actions that give meaning to life. Neither one nor the other can be developed ideally in a highly competitive society that imposes a high work rate and where the other is too often perceived as a potential adversary or competitor. Everyone's energy is all absorbed in this endless competition from which it is dangerous to withdraw for fear of being marginalised or simply downgraded. Where the trap is perfect, is when the overall dissatisfaction that citizens perceive in their life is such that it leads them to want to overconsume in the hope of thus compensating for the absurdity of a life all centred on competition and the increase of their standard of living. Consumption then becomes addictive in the same way as most common drugs: it provides a fleeting satisfaction that quickly gives way to a fundamental dissatisfaction that must be immediately combated through a new consumption that can only be obtained on condition of increasing one's standard of living.[27] The loop is thus closed and we are like rats in a cage exhausting themselves to turn a wheel that nevertheless leaves them perfectly immobile. The ultimate interest of this farce is that of the increase in GDP and therefore, the overall good health of the economy. If everyone exhausts themselves at work and compulsively consumes their income, we enter into a virtuous economic cycle. The one that precisely ensures the growth that our societies need to survive politically and economically, the one that is also the first and fundamental cause of the environmental crisis, hence this terrible dilemma, Charybde and Scylla of the *Sapiens* modern: favour the economy and destroy the planet, protect it and engage in a fatal recession, economically and socially.

2.4 Are Humans Rational in Their Pursuit of Happiness?

Yes and no. No from a certain point of view since they behave, individually and collectively, in a way that is generally detrimental to them and they fall victim to a crude trap—that of the frantic pursuit of wealth and compulsive consumption—with tragic consequences for the planet, and therefore for themselves in the more or less long term. Yes because this behaviour is not strictly speaking irrational in the

[25] Li and Shi (2019).

[26] Oishi et al. (2012).

[27] See Bohler (2021) on the neuronal aspects underlying the processes leading to compensate for the lack of meaning by an overconsumption producing pleasures as ephemeral as unsatisfying.

sense that it obeys a rationality that probably had a real adaptive interest in the distant times of the Palaeolithic and the beginning of the Neolithic, a time when our technology did not allow us to produce and consume en masse nor consequently to destroy the planet. It could very well be that what was adaptive at one time is no longer so today. In this sense, whether our behaviour is rational or not is not so fundamental as psychologists and most economists have admitted and shown that our rationality is limited, far from the ideal of enlightenment, of flawless rationality, theoretically capable of leading us towards individually and collectively ideal behaviours.[28] The question is not so much the rationality of our behaviours and decisions as their adaptive character. To this question, my answer is definitely no. Given recent developments in cognitive psychology and behavioural economics, such an answer is anything but original. We owe in particular to Daniel Kahneman and Amos Tversky, both Nobel laureates in economics, although researchers in mathematical and cognitive psychology, the fascinating demonstration that our decisions, our choices, our behaviours in economic matters and beyond, do not conform to what would result from perfect rationality. On the contrary, we are subject to reasoning and decision biases that frequently lead us astray and lead us to make non-optimal decisions. The list of these biases is significant and I will discuss some of them later to explain our inability to make the right decisions in terms of ecological crisis. For the moment, I will limit myself to presenting a few additional examples of behaviours or decisions supposed to bring us more well-being and happiness and which, in reality, take us away from it.

In an experiment,[29] the well-being level of employees was measured before and 2 months after receiving a bonus that they could use as they wished. Very clearly, those who used this bonus to benefit others (in the form of gifts, donations…) saw their level of well-being increase more than those who used the bonus for themselves as if giving made people happy. However, we cannot exclude that this difference comes from the fact that happy people are more capable of altruism than others. To answer this objection, a second experiment was conducted. This time, some employees were strongly pressured to spend their bonus selfishly while others were almost forced to proceed altruistically. The same result was replicated: those who gave were the most satisfied. Following these two experiments, the authors interviewed employees who had not participated in these experiments. They asked them which among the subjects of this last experiment were the most satisfied with the use they made of the bonus they had been granted: those who spent their money on others or those who benefited personally? Most of the subjects were wrong in their prediction: they wrongly indicated that the happiest were those who had been able to use the bonus for themselves. These results are interesting in two ways. First, they reveal that giving or being generous is more likely to raise the overall level of satisfaction and happiness than hoarding additional gains. This result has been reproduced in numerous experiments and confirmed by several surveys. In reality,

[28] Slovic and Tversky (1974).

[29] Dunn et al. (2008).

the selfish behaviour necessarily associated with competition for the hoarding of more wealth, the very behaviour we adopt in the hope of improving our standard of living and the happiness that is supposed to result from it, is perfectly counterproductive. At the very least, a study conducted among American citizens showed that it was negatively correlated with the level of happiness while altruistic behaviour was positively correlated. This strong link between altruism and happiness was identified in a study covering 139 countries of very different wealth levels[30] and the most comprehensive statistical analyses show the existence of a double causality: being happy leads to being more generous and being generous leads to more happiness. Another example, perhaps less counter-intuitive, concerns the relationship between the level of trust given to others and the level of happiness. Much more than the level of wealth and the associated possibility of consuming at will, the trust that can be given to others plays a decisive role in increasing the feeling of well-being.[31] To put it bluntly, having trust in the people in your environment is more likely to make you happy than having a high standard of living. Yet again, if given the choice between doubling your income and feeling more confident with other citizens, most would choose to double their income.

These two examples are very instructive for three reasons. The first is the existence of an obvious contradiction between the basic postulate of classical economics according to which the increase in wealth leads to an increase in well-being. In reality, altruism and the trust given to others contribute much more to individual well-being.

The second appears to be a paradox. Selfishness and the pursuit of one's personal interest are considered by orthodox economists as the main fuel of a healthy economy. This is the very principle of the old yet still very popular theory of the invisible hand so finely developed by Adam Smith. According to this theory, it should be recalled, the wealth of nations, their harmonious development and the common good result from the selfish behaviour of all economic agents. In other words, the pursuit of personal interest, and therefore the organisation of free competition between economic agents, is necessary for the proper functioning of the economy. However, we have seen that selfishness and the loss of trust given to others, necessary consequences of free competition even regulated by institutions, tend to decrease the level of happiness. To put it simply, it is by seeking our own personal interest that we can enrich ourselves and contribute to the enrichment of the nation, but in doing so, we are in the worst dispositions to be happy. Could there be a global contradiction between the implicit injunctions underlying liberal economics in terms of individual behaviours and those that are likely to make us happy?

The last reason is of great theoretical and political importance. If our economy is called liberal, it is among other things because it places above all, the individual freedom to undertake, to create, to consume… As well as of course to exploit to the hilt other people if it results in a global increase of wealth. The countries that are

---

[30] Aknin et al. (2013).

[31] Lane (2017).

most clearly at the origin of economic liberalism, the USA and the UK, are also the most resistant to all reforms that could lead to a reduction of this fundamental freedom, which freedom has also been constitutive of the democratic revolution that marked the century of enlightenment. In a much more contemporary and concrete way, when the tobacco, alcohol, sugary drinks, junk food and other consumer products harmful to health have been accused of enriching themselves at the expense of consumers, the alibi they have opposed to their detractors through their lawyers was always that of the citizens' freedom to consume or not to consume their products. It must be recognised, despite the terrible damage caused by these companies, that the defence of abused consumers is not as easy as it seems. The problem is actually very complex and would require a critique of the theoretical foundations of liberalism, notably its philosophical conception of free will. Clearly, the studies carried out over the last 50 years at least agree on the following point: the behaviours that we believe are likely to make us happy often lead to us becoming unhappy. We often make choices that are detrimental to us and to suggest that we make them consciously and with total free will is a worn-out fiction that has been largely disproved by the work of psychologists and neuroscientists.[32] Our daily and exhausting battle for personal enrichment, which seems to be the key to our possible happiness, is largely a lost cause: such is the lesson of the Easterlin paradox. The most extraordinary example in this respect is probably that of gambling. Hundreds of millions of poor humans regularly part with their meagre savings by playing the lottery or some other game of chance in the hope of winning the millions that, they think, would make them permanently happy. In reality, winning the lottery most often leads to a sometimes major deterioration in the sense of well-being[33] to the point where help is often needed to deal with the psychological difficulties encountered by the winners. It is generally the deterioration of social ties and the confrontation with emptiness or meaninglessness that contribute to the decrease in the level of happiness. The example of altruism is from this point of view very interesting. Most of the people questioned imagine that personal (selfish) use of a surplus of money will contribute more to happiness than the generous action of benefiting others. Yet it is the opposite that is observed. So finally there is a remarkable convergence between one of the pillars of economic liberalism—the pursuit of personal interest—and common sense. Economic liberalism has been built on a basic and apparently erroneous belief. This convergence of course makes the system extremely robust, but above all leads to strongly questioning the ability we would individually have to make conscious choices as if we had perfect free will. From then on, what credit can be given to this fundamental freedom that should be preserved at all costs if, unbeknownst to them, people who believe themselves to be free are in reality constrained by multiple determinisms of which they are not aware, as Spinoza had formulated in a much more elegant way in his time? How to manage this freedom if it contrib-

[32] See Ripoll (2020) for a psychological, neuroscientific and philosophical analysis of the concept of free will.

[33] Smith and Razzell (1975).

utes moreover to the exploitation of people by other people and to the destruction of the common good that is our environment? In fact, this is one of the basic assumptions of liberalism that collapses. Not only do individuals not adopt rational behaviours, at least adaptive ones, leading to more individual happiness, but moreover, they participate through their frantic search for personal interest and their compulsive consumption in the destruction of the common good necessary for our survival: the Earth. We are once again confronted with the dangers of navigating in the tumultuous waters of Charybdis and Scylla. Without restriction of individual freedom and of the free competition engaged in by states as well as individuals, the chances of preserving a viable planet seem slim. But what would a society be like if it managed to save our planet through constraint and deprivation of freedoms? I don't know which of these two dangers is worse, but it is certain that the later we react to the environmental crisis, the greater the risk that political authoritarianism will be imposed to deal with it, thus marking the end of the era of democracies. Neither one nor the other is desirable.

2.5 The Relationship Between Wealth Level, Consumption Level and Happiness Level in the Light of Greek Philosophy

Everything that recent studies conducted in psychology and economics have rigorously and empirically revealed was clearly anticipated by classical Greek and Roman philosophers. These distinguished two forms of happiness. The first is akin to pleasure, to immediate satisfaction, often linked to material satisfaction or enjoyment. The second, eudaimonia—eudemonia in Greek—is rather a stable state of inner peace, a form of bliss and spiritual peace, a form of harmony between one's own ethical values and one's behaviour. Eudaimonia taken to its extreme could generate an absence of desire since the latter is born from lack and frustration incompatible with eudaimonia.

These two conceptions of happiness opposed Plato and the sophists. For Plato, true pleasures are stable and do not renew as soon as they are satisfied. They reveal an agreement between the individual and the universe and between the individual and the society in which he lives. True happiness cannot primarily depend on material contingencies. It involves a search for spiritual satisfactions guided by reason, which reason must play a major role in an organisation of city life allowing the greatest number to achieve this stable state of inner satisfaction. Conversely, for the sophist Protagoras, Man is the measure of all things and the individual is more valuable than society. Since Man is the measure of all things, good coincides with pleasure so that the frantic search for pleasure, immediate pleasure, is a perfectly reasonable objective. There is no need to interpose between desire and its satisfaction because a happy life is a life where everyone has the means to satisfy their own desires. Therefore, Man's vocation is to dominate nature through technique, to transform it, to adapt it so that it satisfies his infinite needs. In his dialogue with Socrates, the sophist Callicles claims that happiness can only be achieved if nothing hinders the individual satisfaction of desires. Plato then amusingly compares the hedonism of the sophists to the plover, this mythical bird that excretes as quickly as it eats and which cannot stop eating without risk of dying. Plato stigmatises the state of non-satiety of a human soul incapable of managing pleasure. The soul subjected to the attraction of pleasure would be like a bottomless container that empties indefinitely as it is filled, leading the individual into a vortex from which he cannot escape. Plato's intuition has since been confirmed by neuroscientists. The neurotransmitters that contribute to immediate pleasure are not the same as those that contribute to happiness. Those that participate in pleasure intervene in most addictions and addictive dependence is incompatible with this perennial form of happiness corresponding to the concept of eudaimonia.[34]

There is a striking proximity between this old philosophical analysis and the analysis we have made of the Easterlin paradox. It clearly appears that the very foundation of classical economics and its utilitarian approach was built on the position of the pre-Socratics and the sophists. Each economic agent must devote himself to the satisfaction of individual and commercial pleasures. Only satiety can end this objective, but as the level of satiety increases in proportion to the wealth acquired, new desires will always appear, keeping Man in an endless quest for new satisfactions. This is in fact one of the engines of economic growth on which our societies are catastrophically dependent: keeping the human in a state of eternal desire and chronic frustration to push him to consume ever more. Nothing should hinder the free search for new pleasures because from this individual search will emerge a society of opulence—the famous invisible hand—, which is why the measure of GDP, although contested on all sides,[35] remains, almost a century after its invention, the most used criterion to evaluate the prosperity of a nation. Only one essential element is missing for this infernal cycle to never end: innovation and technological mastery that the sophists had already identified as necessary for the creation of new desires and increased possibilities to satisfy them.

2.6 Was Plato Right?

In a certain sense yes but not completely. He was right in the sense that empirical studies conducted on happiness have clearly shown that the enjoyment of material goods and the compulsion to consumption do not produce much additional happiness. He had therefore perfectly identified the absurdity of this bottomless

[34] See Bohler for a detailed analysis of this Phenomenon and Bronner (2021) for its implications in terms of processing information available on the Internet.

[35] Costanza et al. (2014, 2014); Stiglitz et al. (2009); Fioramonti (2013).

psychological hole which ultimately has been the main driver of human activity and creativity, of its economy, probably of its wars but of course today of the destruction of the planet. It is particularly interesting to note that this analysis is 2300 years old, at a time so different from ours, probably indicative of the fact that this determining trait of human psychology is very deeply inscribed in our history, and even our biological history. It could very well be that all the sophistication and complexity of our economies, and more broadly of our society, is based on an archaic psychological trait that dates back to our prehistory. He was also right because empirical research on happiness has revealed that the factors most likely to increase it are hardly related to the enjoyment of material and commercial goods.[36] Thus, among the most powerful contributors to happiness, we find: autonomy, social capital, the level of democracy, citizens' rights, the quality of justice, the level of social justice, job satisfaction, civic engagement, the possibility of living in a natural environment, adherence to strong forms of spirituality, physical activity, the quality of the love relationship, the feeling of acting in accordance with one's ideal of life… The list is long and perfectly heterogeneous, the common point of this list being that these elements are neither linked to a particular market value nor to a classic act of consumption.[37] This is not a speculative list but elements from rigorous empirical studies revealing that people who meet these criteria are on average the happiest.

Where Plato's analysis is problematic is that one can question the compatibility of eudemonism with the reality of human nature. Eudemonism is not unique to Greek philosophy. It is found notably in Stoicism and Buddhism and their quest for a mental state not subject to material contingencies. But isn't it ultimately the expression of a magnificent utopia wishing to make Men Gods, that is, beings who are self-sufficient? Beings for whom completeness is so perfect that desire, envy, greed, the need for power and domination have been completely eclipsed. By setting the bar too high, we risk never reaching it. Worse still, the die-hard idealist, the one who ignores or pretends to ignore the archaic forces that animate humans is probably very poorly prepared to organise a society capable of tempering his greed and thus stopping his fast and mad race towards a major ecological catastrophe. The best we can do is to become aware of the underground forces that animate us. This is probably the only way we have to be able to control them.

References

Aknin, L. B., Barrington-Leigh, C. P., Dunn, E. W., Helliwell, J. F., Burns, J., Biswas-Diener, R., Kemeza, I., Nyende, P., Ashton-James, C. E., & Norton, M. I. (2013). Prosocial spending and well-being: Crosscultural evidence for a psychological universal. *Journal of Personality and Social Psychology, 104*(4), 635–652.

Benjamin, B. D., Heffetz, O., Kimball, M. S., & Rees-Jones, A. (2012). What do you think would make you happier? What do you think you would choose? *American Economic Review, 102*(5), 2083–2110.

[36] Diener and Seligman (2002).

[37] Uchida and Oishi (2016); Ryan and Deci (2001).

Bohler, S. (2021). *Où est le sens ? les découvertes sur notre cerveau qui changent l'avenir de notre civilisation*. Paris, Pocket.

Bowerman, T. (2014). How much is too much? A public opinion research perspective. *Sustainability: Science, Practice, & Policy, 10*(1), 14–28.

Brickman, P., Coates, D., & Janoff-Bulman, R. (1978). Lottery winners and accident victims: Is happiness relative? *Journal of Personality and Social Psychology, 36*, 917–927.

Bronner, G. (2021). *L'Apocalypse cognitive*. Paris, PUF.

Carlsson, F. (2007). *Do you enjoy having more than others? Survey evidence of positional*. Working papers in Economics no 100.

Caunt, B. S., Franklin, J., Brodaty, N. E., & Brodaty, H. (2013). Exploring the causes of subjective well-being: A content analysis of peoples' recipes for long-term happiness. *Journal of Happiness Studies, 14*, 475–499.

Costanza, R., Kubiszewski, I., Giovannini, E., Lovins, H., McGlade, J., Pickett, K. E., Ragnarsdóttir, K. V., Roberts, D., De Vogli, R., & Wilkinson, R. (2014). Time to leave GDP behind. *Nature, 505*, 283–285.

Costanza, R., McGlade, J., Lovins, H., & Kubiszewski, I. (2014). An overarching goal for the UN sustainable development goals. *Solutions, 5*(4), 13.

De Neve, J. E., Ward, G. W., De Keulenaer, F., Van Landeghem, W., Kavetsos, G., & Norton, M. I. (2015). *The asymmetric experience of positive and N, G., negative economic growth: Global evidence using subjective well-being data*., IZA Discussion Papers, No. 8914. Institute for the Study of Labour (IZA).

Diener, E., & Seligman, M. E. (2002). Very happy people. *Psychological Science, 13*, 81–84.

Diener, E., Tay, L., & Oishi, S. (2013). Rising income and the subjective well-being of nations. *Journal of Personality & Social Psychology, 104*, 267–276.

Diener, E., et al. (2018). Advances and open questions in the science of subjective well-being. *Collabra: Psychology, 4*(1), 15. https://doi.org/10.1525/collabra

Dunn, E. W., Aknin, L. B., & Norton, M. I. (2008). Spending money on others promotes happiness. *Science, 319*, 1687–1688.

Easterlin, R. A. (1974). Does economic growth improve the human lot? Some empirical evidence. *Nations and Households in Economic Growth, 89*, 89–125.

Easterlin, R. A. (2017). Paradox lost? *Review of Behavioural Economics, 4*(4), 311–339.

Easterlin, R. A., McVey, L. A., Switek, M., Sawangfa, O., & Smith Zweig, J. (2010). The happiness–income paradox revisited. *PNAS, 107*(52), 22463–22468.

Fioramonti, L. (2013). *Gross domestic problem: The politics behind the world's most powerful number*. Zed Books.

Griffin, D., & Tversky, A. (1992). The weighing of evidence and the determinants of confidence. *Cognitive Psychology, 24*, 411–435.

Haushofer, J., & Shapiro, J. (2015). *Household response to income changes: Evidence from an unconditional cash transfer program in Kenya*. Working paper. The Busara Center for Behavioural Economics.

Hsee, C. K., & Hastie, R. (2006). Decision and experience: Why we don't choose what makes us happy? *Trends in Cognitive Sciences, 10*, 31–37.

Kahneman, D., & Deaton, A. (2010). High income improves evaluation of life but not emotional well-being. *PNAS, 107*(38), 16489–16493.

Kanheman, D., & Tverski, A. (1979). Prospect theory: An analysis of decisions under risk. *Econometrica, 47*(2), 263–291.

Kerschner, C. (2010). Economic de-growth vs. steady-state economy. *Journal of Cleaner Production, 18*, 544–551.

Lane, T. (2017). How does happiness relate to economic behaviour? A review of the literature. *Journal of Behavioral and Experimental Economics, 68*(2017), 62–78.

Layard, R., Mayraz, G., & Nickell, S. (2010). Does relative income matter? Are the critics right? In E. Diener, J. Helliwell, & D. Kahneman (Eds.), *International differences in well-being*. Oxford University Press.

Li, L., & Shi, L. (2019). Economic growth and subjective well-being: Analysing the formative mechanism of Easterlin Paradox. *The Journal of Chinese Sociology, 6*, 1. https://doi.org/10.1186/s40711-018-0090-9

Lucas, R. E., Clark, A. E., Georgellis, Y., & Diener, E. (2004). Unemployment alters the set point for life satisfaction. *Psychological Science, 15*, 8–13.

Lucas, R., & Donnellan, M. (2007). How stable is happiness? Using the STARTS model to estimate the stability of life satisfaction. *Journal of Research in Personality, 41*(5), 1091–1098.

Luttmer, E. F. (2005). Neighbours as negatives: Relative earnings and well-being. *Quarterly Journal of Economics, 120*, 963–1002.

Medvec, V. H., Madey, S. F., & Gilovich, T. (1995). When less is more: Counterfactual thinking and satisfaction among Olympic medalists. *Journal of Personality and Social Psychology, 69*, 603–610.

Oishi, S., Schimmack, U., & Diener, E. (2012). Progressive taxation and the subjective well-being of nations. *Psychological Science, 23*(1), 86–92.

Ripoll, T. (2020). *Pourquoi croit-on ? Psychologie des croyances*. Auxerre, Sciences Humaines.

Ryan, R. M., & Deci, E. L. (2001). On happiness and human potentials: A review of research on hedonic and eudaimonic well-being. *Annual Review of Psychology, 52*, 141–166.

Slovic, A., & Tversky, A. (1974). Who accepts Savages's axiom? *Behavioural Science, 19*(6), 368–373.

Smith, S., & Razzell, P. (1975). *The pools' winners*. Calibon.

Solnick, S. J., & Hemenway, D. (1998). Is more always better? A survey on positional concerns. *Journal of Economic Behaviour & Organisation, 37*, 373–383.

Stiglitz, J., Sen, A., & Fitoussi, J.-P. (2009). Report of the Commission on the Measurement of Economic Performance and Social Progress.

Uchida, Y., & Oishi, S. (2016). The happiness of individuals and the collective. *Japanese Psychological Research, 58*, 125–141.

Chapter 3
Why Can't We Stop Consuming?

Abstract The human impact on the planet is attributable to our propensity for excessive consumption. But what are the determinants that compel us to consume excessively? There are many. Of course, we can mention the well-known characteristics of the so-called consumer society and the ingenuity of advertising campaigns. But it would be very naive to believe that we consume simply because society has implemented incredible subterfuges to trap us. In reality, there are powerful forces, biologically ingrained in our brain, that make us almost pathological consumers.

There is nothing more commonplace than consuming. At almost every moment of our lives, we consume. Even when we sleep, we consume: the bed, the sheets, the blanket and the heating necessary for a comfortable sleep are indeed consumer goods, products or services that have resulted from manufacturing work carried out from the exploitation of natural resources. We spend most of our lives consuming because life is not possible without an act, however minimal, of consumption. Our relationship to the world is therefore largely a relationship of consumption. Society itself can be considered as a gigantic organism with a metabolism that allows it to consume—assimilate, destroy and reject waste—in order to develop, renew and evolve. The whole question is to identify the level of metabolism compatible with the resources we draw from the Earth's environment. The fact is that there would be no environmental problem if our society's consumption was reasonable. But it is not for multiple reasons, some of which were outlined in the previous chapter. There is nothing surprising about this as the number of factors that lead us to consume excessively are numerous.

T. Ripoll, *Human Psychology and the Destruction of the Planet*, Anthropocene – Humanities and Social Sciences, https://doi.org/10.1007/978-3-032-02080-2_3

3.1 Invisible Consumption

We often consume without being aware of it, often simply due to a lack of information or scientific culture. Here are a few examples. A few years ago, when computing and digital exchanges were beginning to develop, I remember a discussion I had with one of my colleagues, who believed that with the explosion of digital technology, our energy consumption would decrease considerably. As everyone knows, it has since increased significantly, largely due to digital technology, which is now estimated to account for 3–4% of the planet's total consumption and about 10% of global electricity consumption. Watching a film in streaming, storing a few thousand photos on a cloud or simply leaving your mobile phone on day and night actively contributes to the degradation of the environment.

It is only a few years ago that we became aware of the environmental impact of meat consumption. Less than 20 years ago, hardly anyone thought that eating meat contributed to the destruction of the planet. When we eat a steak, our carbon footprint is the same as that resulting from a 100 km car journey, not to mention the thousands of hectares of forest and fauna that we indirectly destroy to satisfy our taste buds alone. As often, our brain is unable to grasp the problem because our cognitive horizon is limited both by its inability to grasp what is happening far from us and by the enormity of the energy flows at stake. The fact that we learn that 65 billion farm animals are killed each year, 2000 per second, and that hundreds of thousands of wild animals are eliminated due to the destruction of their biotope, should have an impact on anyone who is supposed to be sensitive to the environmental cause. But the impact is often short-lived because these figures are beyond us, or at least beyond the understanding capabilities of our brain.

The case of Vegans or vegetarians who raise dogs or cats is even more paradoxical. At a Vegan demonstration, I saw several protesters holding a dog on a leash. When I questioned them about the inconsistency between feeding a dog or cat with meat while they themselves refused to consume it to preserve the environment, some looked at me with astonishment, puzzled by the question I asked them. Others, undoubtedly very ignorant about ecology, retorted that dogs are animals, that they are therefore part of nature and that it is natural for them to eat meat. They were not even aware that dogs, like cats, are only human extensions that we have made for our own interest. The environmental impact of dogs and cats is considerable. By simply feeding them, we generate a carbon footprint equivalent to that generated by 13 million cars.[1] In the USA, the food of cats and dogs represents 30% of the environmental impact of agriculture. As for the loss of biodiversity, it is considerable and has led to the extinction of endemic species especially in islands or recently colonised lands like Australia or New Zealand. 63 animal species have permanently disappeared in 500 years because humans have raised or introduced, in areas where they had no business being, the pets they cherish so much.[2] In France, our 12 million cats are insatiable predators. They prey on 324 million wild animals, some of which

[1] Okin (2017).

[2] Crooks and Soule (1999).

are endangered. Unlike respectful hunters, they do not differentiate between a protected species and one that is not.[3]

All these examples, among thousands of others possible, illustrate the difficulty of becoming aware of significant consumption acts that we do not perceive as such. This type of problem is nevertheless not insurmountable provided that the public is informed or accepts to inform themselves and that denial does not prevail over the acceptance of indisputable data: how many of us raise a cat while lamenting the loss of biodiversity? Becoming aware that we consume during activities not identified as consuming is the first step, and probably the easiest, towards reducing global consumption.

3.2 The Norm in Terms of Consumption or "Normal" Consumption

In all countries, and primarily in Western countries, the level of consumption per capita has grown almost exponentially in recent centuries. Between the 1950s and 2000s, in England for example, the average level of consumption increased by 100%.[4] In the field of leisure, the increase is 400%, in that of household objects, it is 385%, 341% in the field of communication, 200% for clothing, only 29% for food and 14% for culture. It is likely that people who evaluate themselves today as frugal are actually very large consumers compared to what their grandparents were. We are in the reign of hyperconsumption. We are so immersed in this dysfunction that we can no longer even grasp its absurdity and danger, like these chronic alcoholics who are no longer even aware that they drink outrageously or these obese people who wonder about a weight gain that does not seem related to their diet. The housing sector is symptomatic of a staggering increase in consumption that nevertheless ends up seeming natural and legitimate to us. At the beginning of the century, a family of five people frequently lived in a 60 m^2 area. Today, the average area is 91 m^2, often occupied by 1 or 2 people. When I was a child, a 200 m^2 house was a huge mansion. Today, 250 m^2 houses don't even attract attention and most of their owners often have a second home. In the USA, the size of individual houses has increased by 65% since 1970 even though the number of its inhabitants has decreased.[5] Swimming pools were luxury items that could be counted on one hand in a medium-sized agglomeration in France. There are now nearly three million of them, a rapidly increasing figure. The most incredible thing is that this increase in living space, associated with a concomitant increase in all sorts of household objects, has generated multiple problems for their owners: financial, technical, logistical, temporal—it takes time to take care of pharaonic housing—and even psychological. One in ten American households now rents a storage space to store the useless bric-a-brac that overflows from the available space in their home. An

---

[3] Loss et al. (2012).

[4] Jackson and Marks (1999).

[5] Samuelson (2005).

industry was born as private storage became the fastest growing sector in real estate over the past two decades, with the number of premises dedicated to the storage of unnecessary items increasing by 81% from 2000 to 2006. According to an Australian study, four out of ten people feel anxious, guilty or depressed due to the clutter in their homes.[6] Some feel trapped, oppressed by what they own and admit to having a room they would be ashamed to show. The need to own, to accumulate is so compelling that professionals are now trained to help those psychologically unable to manage their Aladdin's cave. According to a study presented by Clive Hamilton,[7] Australians spend more than 1200 dollars per household on consumer goods they never use, slightly more than what the government spends on universities.

The most astonishing thing about this hyperconsumption, even when people are aware that it is the cause of a waste of time, money and psychological difficulties, is that it is not experienced as hyperconsumption, or even sometimes as consumption. For those who are victims of it, it is nothing more or less than a usual practice, a norm, a standard, sometimes a way of life, or even an ecologically virtuous behaviour. This is particularly the case with these charming flea markets that have proliferated on the outskirts of large cities and can thus give their followers the feeling of participating in an ecological recycling that is both ethical and poetic. When people who can no longer manage the clutter they have accumulated call in professional organisers, they do not conclude that they have accumulated too much and therefore consumed too much. They simply think they are incapable of managing their sad treasure. For most, they are not even aware that such accumulations represent, on a global scale, a significant cause of the destruction of the planet. These usual consumer behaviours, which are no longer even aware of consuming, have been called practices.[8] In fact, consumers no longer perceive themselves as consumers, but as citizens following usual practices, behavioural routines, which are part of a social norm that is not excessive since it is accepted by all. We are far from ostentatious consumption, which is carried out to show off social status and often involves acquiring luxury items. Among the examples of practices that have been precisely studied, we can mention: taking a shower twice a day, air conditioning in cars and homes, washing our clothes daily, flying abroad for holidays, lighting up our garden or the entrance to our house day and night, surfing the internet for hours on end... What is extraordinary about practices is that once they have been acquired, they become so integrated into our lifestyle that we completely lose awareness of what they represent. It is likely that you would consider it absolutely regressive to question some of the ones I have just listed, especially those that seem to be related to our lifestyle. This is a clear sign of their perfect and deep integration into our way of life to the point that it is no longer possible to imagine getting rid of them without admitting an unacceptable deterioration of a comfort that we consider minimal. Very slowly, collective usage does its work and the force of social mimicry quickly convinces us that what was not even imaginable yesterday has become a necessity today. No one could have imagined 50 years ago that it would seem necessary to us

---

[6] Fear (2008).

[7] Hamilton (2010).

[8] Shove (2003); Ropke (1999).

today to take a shower two or three times a day. The very notion of "clean" has completely evolved and a new norm has imposed itself to the point that we would almost be disgusted by the dirtiness of the clothes worn by our close ancestors, which they of course found very clean. Even our perception of the whiteness of a fabric has incredibly evolved. The "white" of our grandparents or great-grandparents was not the same white as ours! Roughly speaking, in 2000, we washed our clothes 3 times more frequently than in the 1950s[9] and Americans, at the same time, always washed their clothes twice as much as Europeans. We are always quick to be surprised or to criticise the excess consumption typical of the United States, but a few years are usually enough to integrate their practice and consider as normal what seemed superfluous a few years earlier. This phenomenon is so powerful that it even seems to impact our visual perception. If you are a car enthusiast and you get to see a comfortable saloon car that you knew in the 80 s, it will probably seem quite tiny to you. In the field of boating, in the 80 s, a 10-metre yacht gave an incredible impression of power, safety and comfort to the sailors who coveted it. Today, such boats look so frail that we struggle to understand what we felt at the time.

The concept of "comfort zone" perfectly illustrates this phenomenon. In terms of temperature, for example, the commercial communication of many companies presents the concept of comfort zone as if it were an objective concept: the optimal zone for the physiology of every human being. All the skill and perversity the purpose of such communication is to persuade consumers of the existence of a standard that would no longer be social and therefore relative, but absolute and biologically determined. Of course, this is not the case. In the second half of the twentieth century, the so-called normal temperature in an apartment in the USA was 22 degrees. Below this, it was considered an area of discomfort as it seemed absurd to have to wear a jumper at home. Today, for most motorists, it is inconceivable not to have air conditioning in their car. This didn't even occur to European drivers 30 years ago. Similarly, the home air conditioning market is literally exploding in France (almost 30 years after the USA) because the high temperature of our comfort zone is now estimated at 23 degrees as if humans could only live well in a temperate and spring-like climate, when the temperature fluctuates between 20 and 23 degrees. Thus, incidentally, we regularly integrate new practices so constitutive of our daily life that it becomes almost impossible to challenge them or question their real necessity. They are no longer even identified as acts of consumption even though they contribute massively to the destruction of the planet.

3.3 The Norm in Terms of Practice and Its Systemic Effect

The norms in terms of practice can evolve, which signifies their eminently cultural dimension. Sometimes they are revised downwards: this is quite rare. This is the case today, for example, with what is considered normal meat consumption. But

[9] Biermeyer (2001).

generally, these norms are always revised upwards or, even worse, new practices appear, generally linked to a technological innovation. What is very powerful and constraining is their systemic character in the sense that the establishment of a practice has a cascading effect on entire sections of society, bringing with it the establishment of other practices from which it is difficult to escape. This is typically the case with mobile phones. No one can do without them because the very organisation of our life in its most ordinary aspects has been modified by their presence. To economise on them is to marginalise oneself to the point of risking exclusion. I remember the extreme case of a young teenager who said she would rather be deprived of food for a day than her mobile phone and who threatened to commit suicide if her parents persisted in their decision to take it away from her. The car is another example. It is not a question of denying its usefulness but simply of noting that architecture and urban planning have been completely rethought because of its presence in the lives of all Westerners. The existence of large shopping centres on the outskirts of cities could only develop with the generalisation of the car among the middle classes. The same is true of the expansion of large suburbs often located tens of kilometres from urban and economic centres. Even if some people would like to do without their car, it is often impossible because our societies have been designed with it in mind. The case of air conditioning in commercial and professional premises, which is now the norm, has led to the disappearance of an ancestral tradition: the siesta, once necessary to rest the body during the hottest hours of the day. The siesta is gradually disappearing from the entire planet. It was even banned by the Mexican government in 1999, considering that the generalisation of air conditioning could not justify this temporary stoppage of economic activity. The fact that practices form a system makes their questioning almost impossible. Even when they appear harmless to us, they are so deeply articulated with a multitude of other practices that it is not obvious to evolve or challenge them.

3.4 Malleability of Norms and Ratchet Effect

The ratchet effect has been well analysed by sociologist Elisabeth Shove. It is a basic psychological mechanism that has a strong impact at the societal level. When a new level of comfort is reached, when a new technology becomes accessible or when there is any improvement in our living conditions, there is initially a positive perception. Then a rapid adaptation leads to the disappearance of this feeling of positivity, heralding the establishment of a new norm. The "tragic" part of this story is that once the new state has become the norm, it becomes psychologically impossible to go back. If we get used to a temperature of 20 degrees in the summer in our home due to the use of air conditioning, it will no longer be bearable to sweat even a little at home, which was not a particular problem for previous generations because the norm was not the same. This mechanism, close to the phenomenon of hedonic habituation described in the chapter on happiness, also has a biological, and even neuronal basis. Our nervous system is programmed to perceive any pleasant (or unpleasant) environmental change. This is precisely what allows it to evolve and actively seek the environments that are most favourable to it. As soon as an improvement has been perceived, it generates a well-being whose neuronal correlate can be identified very precisely. But the effect of this improvement on our neuronal activity, and therefore on our subjective perception of pleasure, is temporary. It corresponds to the time necessary for hedonic habituation to exert its action. For example, if monkeys are given fruit juice, an increase in the activity of neurons in the substantia nigra and the tegmental area is observed. These neurons play an important role in the production of dopamine, a neurotransmitter closely related to our sensation of pleasure. If the amount of juice is doubled, the neuronal activity doubles as well. However, if the same amount of juice is given every day, the neuronal response decreases until it disappears completely. This is the neuronal basis of the ratchet effect. Quantity of juice, we observe that the activation level goes from three discharges per second to eighty. But if we maintain this quantity, it only takes a few minutes for the level of neuronal activation to return to the initial level. The problem is that if the maintenance of this improvement no longer brings satisfaction once hedonic habituation has done its job, its disappearance generates in return a state of discomfort and frustration far greater than what we felt before this comfort improvement occurred. Thus, the ratchet effect makes a loss, even if minimal and objectively unimportant, of our comfort level extremely difficult. This phenomenon is further accentuated by the weight of the social group to which we belong. On the one hand, the very mimetic dimension of our psyche—the herd instinct—constrains us more than we imagine to adopt the practices of the group. On the other hand, giving up any element of comfort when the members of one's own group still benefit from it generates an almost unbearable difficulty giving the ratchet effect its societal, mechanical and almost deterministic dimension. Thus, without external constraint, the level of consumption can only increase, the "always more" of our societies is only the consequence of biological and sociological determinisms that condition us to consume more and more.

There is nevertheless a glimmer of hope in this story: it is that social determinisms can either reinforce biological and psychological determinisms, or annihilate them. Overall, the expansion of our societies, the unchallenged domination of the economy and Western culture, the generalisation of a societal model based on the values of work, desire, competition, consumption, has consisted in reinforcing, and thus exploiting, the psychological mechanisms that we have just described. Regardless of the sophistication of our society, it relies and is based on basic biological and psychological determinisms. But society can also exert pressure which goes against this fundamental biological force that incites us to seek always more comfort and pleasure. Indeed, humans are social animals who always try to individualise and display themselves through messages that are favourable to them or, at the very least, that are in line with the image they want to give of themselves. If tobacco consumption has decreased in France, it is as much due to the negative image it now conveys as to the awareness of its danger. If many senior executives now drive modest Dacias rather than powerful and comfortable SUVs, it's because the message they transmit by driving such vehicles is no longer at all in line with the message they wish to communicate to others. This is a path that needs to be exploited. Only society, through the pressure it can exert on the biological and archaic forces that determine us to consume more and more, is capable of avoiding the environmental catastrophe. For the moment unfortunately, it is clear that it has exploited and outrageously reinforced them.

3.5 Desire, Pleasure and Consumption

For an organism to evolve, in the Darwinian sense of the term, for an individual to adjust his behaviour towards his environment, for him to learn, for him to find appropriate solutions, nature has "developed" a universal tool of infinite power: the sensation of pleasure and displeasure. It is this particular sensitivity to what feels good and what hurts that is at the origin of the complexity of living beings and their evolution. In fact, from the simple virus through plants to higher mammals and Man, every living being seeks the internal state that gives him the most pleasure even if, of course, the pleasure that a human feels in front of a beautiful sunset or that which results from the sweet words addressed to him by the loved one has many other properties than the "pleasure" that a parched plant feels suddenly watered. In all cases, in very different forms, it is this principle of pleasure and displeasure that contributes to homeostasis, that is to say the possibility of preserving a stable and optimal internal state despite the constant variations of the environment: the origin of pleasure goes back to our most distant biological past. As for desire, it largely constitutes the engine of our behaviours, especially those that are likely to give us the expected and sometimes fantasised pleasure. The fact that this sensitivity to pleasure is so ancient from a phylogenetic point of view explains that the neuronal structures that make it possible are very archaic structures located deep within the brain (notably the striatum, the amygdala, the nucleus accumbens and the entire dopaminergic network). These neuronal structures play an important role in our emotions. While they can largely constrain our decisions, they do not have the ability to reason as such, an ability essentially ensured by more recent cortical structures, notably the frontal and prefrontal cortex. They constitute "our animal part", in relation to what is at the foundation of our affectivity, our drives, all this unconscious and sometimes tyrannical life capable of directing us towards paths that our reason can condemn at the same time. The instinctual force that plays with our rationality by making us glimpse the attainment of a near pleasure is so powerful that it can cleverly enslave our capacity for critical and rational analysis. This is what happens when we deploy treasures of rhetoric and intelligence to justify an absurd behaviour[10] such as our compulsion to consumption. The conscious control we have over these structures is therefore very weak, which is why pleasure escapes so much from our will. We do not desire by deciding to desire. Desire imposes itself

[10] Pronin (2007).

on us more than we control it. We do not enjoy at will, just as we resist very difficultly to a immediate pleasure even when we know it is harmful to us in the long term. This is at the heart of our addictions, whatever they may be, and our compulsive consumption is just a particular case. This dependence on pleasure is so intense that it can lead to irrational behaviours and potentially harmful in the long term. This is the case if two main conditions are simultaneously met: when the source of pleasure is no longer constrained by some physical limitations of our environment and when evolution has not endowed us with the ability to regulate our behaviour towards certain forms of pleasure. In the natural environment of the pre-Neolithic Homo Sapiens, no one had at their disposal an abundance of soda cans, fries laden with lipids, not to mention the thousands of artificial or natural drugs that humans constantly seek and consume. Our natural appetite for these kinds of substances and the fairly universal pleasure that results from their consumption posed little risk in the past given the rarity of these substances in nature. Somewhere, we were programmed to gorge on food when we had the rare opportunity to do so. This allowed us to build up the necessary energy reserves to face times of famine. But when we can gorge ourselves constantly, this programming, which had a real adaptive interest, becomes tragically deadly. Hundreds of millions of humans suffer from obesity, a pandemic of formidable power and difficult to combat because it ultimately requires opposing an immediate source of pleasure that it is not natural for us to defer. It should not be believed that addictions are specifically human. It has thus been observed that rats continued to operate a lever stimulating the neuronal structures causing pleasure even if this was followed by the administration of a painful electric shock. Many animals, including elephants and chimpanzees, get drunk more than reason on overripe fruits loaded with alcohol until they develop dangerous behaviours related to heavy alcoholism. Cats and dogs locked in the confined spaces of an apartment and having infinite food become obese as we become ourselves in similar circumstances. In addition, it has been observed in animals, in situations of stress or when boredom resulted from captivity, that the search for compulsive pleasure through food or drug consumption increased. The same reasons explain that all forms of addiction affect humans more in psychological distress. This distress can result from a situation of stress, boredom, loss of meaning and degraded self-image. Addiction then constitutes the most direct and least effective means in reality to overcome these psychological difficulties. In our Western societies of luxury, opulence and comfort, everything contributes to exacerbating our addictions. The level of psychological insecurity is significant, in relation to the primacy given to inter-individual competition. The loss of meaning is very marked largely in relation to the number of individuals forced to earn their living by accepting a job for which they are at best docile performers, forbidden from any initiative and therefore from any creative dimension. Not to mention, the emotional, social, cultural and sometimes sexual misery that now affects young and old alike. In such a context, all addictions can unfold without limits because in the absence of real satisfactions derived from our life in society as in the absence of events likely to reinforce our self-image, it is necessary to fill this narcissistic chasm with some form of consumption capable of very momentarily raising our minimal level of pleasure. All this regardless of the long-term cost to our health, our budget, our time spent working and of course the preservation of our environment.

3.6 The Psychological Value of Time

The reasons to establish a link between the banal forms of addiction and our compulsion to consumption are numerous. Among these, we have implicitly mentioned above the temporal dimension. The problem with addiction is that the person suffering from it does not give the same value to near time and distant time: the relative and psychological weight of the pleasure generated by an immediate intake of nicotine is immensely greater than the distant prospect of future suffering caused by cancer, respiratory difficulties or cardiovascular disorders. The same is true regarding our behaviour towards the looming environmental disaster. At the individual level, few are the individuals who decide not to succumb to the immediate pleasure of any consumption for real ecological reasons. As for our governments, they often commit to taking the measures recommended by scientists to limit our CO_2 emissions. But as soon as the economic crisis approaches, as is the case with the coronavirus at the time I am writing these lines, the right decisions are postponed since the clearly displayed obsession at the present time is to boost consumption to try to reduce the monstrous debt and avoid economic recession. In a certain way, the environmental crisis is a direct consequence of our extreme sensitivity to the present or near future and a form of psychological devaluation of the more distant future. The psychological value of time has been extensively studied by economists and psychologists. In economic terms, we speak of preference for immediate utility over future utility. In psychology, we talk about temporal discount rate or impatience rate. The joint work of psychologists and economists has clearly challenged the possibility that humans have to make choices and decisions in accordance with an ideal form of rationality that allows them to effectively project into the future. Our rationality is said to be limited, in the sense that our decisions do not manage to optimally adjust to the situations we will face in the future. Adam Smith had already noticed that economic agents have a way of managing the temporal dimension that does not seem optimal in relation to their desire to accumulate wealth. Agents seem to be subject to a form of tyranny of immediate pleasure, making long-term planning and asset management very difficult. In economics, immediate enjoyment seems to easily take precedence over future pleasure.[11] In short, we would generally be more like the grasshopper when considering our medium or short-term future would require us to be more like the ant. Economic behaviour is just another psychological behaviour, the link with addiction is obvious. For a human brain, a few years represent such a large temporality that it is difficult to modify the power of our

[11] Robson (2002).

dependence on immediate pleasure.[12] This cognitive weakness is further reinforced by the fact that the future consequences of our actions, favourable or unfavourable, are generally of a statistical or probabilistic nature and that, again, we are particularly poorly equipped cognitively to process statistical and probabilistic information. Most smokers know that the probability of physically suffering from their addiction in the long term is considerable. But these are still statistics, which moreover are statistics about a distant future. This is very insufficient to hold a candle to their need for immediate pleasure. How to resist the immediate neuronal pleasure caused by a simple cigarette when cancer might only occur in a few years with a probability that is certainly high, but which remains a probability. "Go on, one last one" or "it's not one more that will make a difference" or "at this point, it won't change much" are the typical arguments that the addictive person develops to allow themselves these little suicidal pleasures. These are the same arguments that we develop in relation to our excesses of consumption of all kinds, especially those that contribute to the destruction of our planet.

It is quite easy to measure the discount rate. Experimentally, behavioural finance specialist Richard Thaler,[13] has developed a simple test to evaluate our personal discount rate. For this, a subject is asked what amount would be equivalent to 15 dollars available immediately if this amount were to be granted in 1 month, 1 year and 10 years. The median of the responses is respectively at 20, 50 and 100 dollars. In other words, subjects would be willing on average to defer their profit by 1 month, 1 year and 10 years if the amount granted to them increased by 33%, 233% and 566% respectively. The further away the deadline, the less willing we are to give up an immediate advantage. Conversely, the more a negative effect of a behaviour is likely to manifest itself late, the less capable we are of taking it into account, even when the negative effect is certain. This is a form of cognitive bias that can lead to very bad decisions simply because the hedonic component is at the heart of economic decision-making as well as all human action.[14] Field studies conducted by economists have shown that the propensity to buy on credit in economically dangerous proportions was much higher among households with a psychological profile characterised by a high discount rate.[15] In another study,[16] soldiers were offered the option of receiving a lump sum or a pension spread over a period of 18 years. Although the pension was very economically advantageous and therefore rationally preferable, a minority made this choice.

Why are we so subject to the tyranny of the present or the short term? Sébastien Bohler has analysed this phenomenon based on neuroscientific data suggesting that the neural structures involved in the reward circuit and therefore pleasure exert a greater power than the neural structures dedicated to reasoning and planning. These

[12] Heather and Segal (2013); Henden (2012); Henden et al. (2013).

[13] Thaler (1981).

[14] De la Bruslerie (2012).

[15] Harris and Laibson (2001).

[16] Warner and Pleeter (2001).

ancient neural structures are very sensitive to the present time because only the present time is capable of providing us with the pleasure we all seek. This of course explains why the battles we fight against the small and guilty immediate pleasures of everyday life are often lost in advance. It is important to understand here that the power exerted by the neural structures involved in the reward circuit is not an anomaly of our evolution, nor even a form of pathology (except of course when the addictive dimension becomes too great). Evolutionary psychology provides a very convincing explanation of this rather unbalanced balance of power. In our past as *Sapiens,* let alone in our even more distant past as hominids, the cognitive horizon rarely exceeded that of the day. Even current anthropoid monkeys with their remarkable intelligence are not inclined to think over long periods of time. Their most common goals do not go beyond a few hours. For our Australopithecus ancestors, it was probably the same, especially as their life expectancy was reduced: seeing too far ahead was not very adaptive as it was already so difficult to survive in the short term. When the primary concern is to feed, take refuge, reproduce, a good strategy is to deploy all one's energy and intelligence to solve the most immediate problems, those directly linked to our survival and our ability to pass on life. Implementing a strategy to capture a mammoth to enjoy its flesh on the same day was much more adaptive than projecting oneself at the dawn of one's own death, or even better, into the lives of future generations. Our brain has been programmed to give more importance to immediacy. We can conceive the future much better than current monkeys, but we do it with recent neural structures (the neocortex) which cannot compete with the old neural structures involved in the hedonic and emotional component, closely related to the present time. It is clear that the Neolithic revolution probably began to change the game. The fact of cultivating, preserving, settling down, making reserves and having to organise a complex and highly structured society has forced us more to think about the long term. But the Neolithic was only 10,000 years ago when the evolutionary processes that shaped us took place over millions of years. This is very little in the light of our biological history, too little in any case to provoke a modification of our cognitive system allowing us to rationally process the distant future. Of course, in the old framework of our natural environment, the search for immediate pleasure and the importance of the present time did not pose any real problems, quite the contrary. It is quite different today, on the one hand because our society makes a quantity of immediate pleasures previously absent from our environment accessible, and on the other hand because the level of consumption that results from it is at the origin of the environmental crisis we are experiencing. Not only is it difficult for us to resist these sources of hedonic satisfaction, but we also have great difficulty in concretely representing the long-term consequences of our bulimia: we manage to intellectually conceive the coming ecological disaster, but in reality, it remains very abstract, very theoretical and emotionally inert for our poor brains of *sapiens*, apparently incapable of reacting to the imminent danger.

It is possible that such an analysis may seem terrifying and despairing. If our behaviour, especially our hedonic greed and our inability to think about the future, are so biologically and psychologically determined, is there a solution to the problem of the environmental crisis? Nothing is less obvious, but nothing in this analysis forbids us from hoping. Two important authors have proposed a conception of human psychology compatible with a minimal form of hope.

The first is Freud. Freud admitted that our behaviours are subject to two antagonistic forces: the pleasure principle and the reality principle. The pleasure principle corresponds to the drives, often unconscious, leading us to seek immediate pleasure whatever the cost to ourselves or others. The reality principle is the awareness of the existence of limits, constraints, obstacles to obtaining this pleasure. Taking into account the resistance of reality is necessary for one's own survival and especially for a peaceful life in society. It implies the ability to postpone the expected pleasure, or even to renounce it when immediate satisfaction can produce suffering for oneself or others in the longer term. The reality principle also requires us to get out of the fantasy of omnipotence, of an egotism and selfishness that would then be the mark of a form of psychological immaturity. For Freud, the conflict between these two principles is at the heart of the problems that the individual encounters and, consequently, our society as a whole. However, it is society, the force of reality itself, education and the maturation of the psychic system that can contribute to the reality principle taking precedence over the pleasure principle. This would not imply the abandonment of all forms of hedonic satisfaction, but a search for pleasure constrained and limited by the requirement to take into account reality. Nothing is therefore decided in advance concerning this struggle at the heart of human psyche because if it results from a conflict of biological nature, humans are also beings of culture and the latter is of formidable power, capable in the best case of an effective arbitration and, in the worst case, of reinforcing the dark side of human psychology: the one that lets the pleasure principle and our compulsive search for immediate satisfactions unfold without restraint. The very existence of these two principles—reality and pleasure—is therefore a reason for hope. There are indeed in human nature some reasons to hope because nothing allows us to know which of these principles will take precedence over the other in the management of our societies.

The second author, or rather, the couple of authors, David Kanheman and Amos Tversky, both Nobel laureates in economics although psychologists, are at the origin of a conceptual dichotomy largely confirmed by a considerable number of empirical studies in psychology and economics.[17] These authors have identified two modes of operation of the cognitive system: the intuitive mode and the analytical mode. From a phylogenetic point of view, the intuitive system is undoubtedly the oldest. Everything suggests that our distant Australopithecus ancestors had this same information processing system simply because it has characteristics that give it great adaptive power, especially through its ability to respond quickly to situations in ordinary life. Classically, this system is defined by the following characteristics. It is fast, largely appeals to the memory and our past experiences. It is cognitively inexpensive. It is automatic and weakly involves our reasoning abilities. It is

[17] Kahneman (2011).

extremely sensitive to the emotional dimension and therefore largely determined by affect. It is primarily guided by the search for solutions capable of immediately increasing our level of satisfaction. Overall, this system of thought is a default system, that is, the one we use most naturally and frequently. Its main limitation is that it is the source of many cognitive biases, often a denial of part of reality and it leads to inappropriate decisions when a fine and complete analysis of reality is required to behave optimally. Conversely, the analytical system is very recent on the phylogenetic scale and very likely linked to the emergence of language. Its recentness on the evolutionary scale means that we use it less frequently and less spontaneously. Therefore, it does not activate automatically and requires a deliberate and voluntary intention, costly for the subject. This system is slower and requires significant cognitive effort. It heavily solicits our working memory, that is, our ability to keep the information necessary for decision-making conscious. It uses reasoning and proceeds by logical inferences. It is little dependent on the emotional dimension. It is acquired through education and generally develops in proportion to the level of study reached.

One could say that a major contribution of these authors is to have revealed the prevalence of the intuitive system over the analytical system and its consequences in the decisions we make. If the Nobel Prize in economics was awarded to these psychologists, it is precisely because they have revealed, through remarkable empirical studies, that the economic agents that we are often make aberrant decisions, at least non-optimal ones leading us to make choices that are unfavourable to us. This is systematically the case when the situations to be dealt with become excessively complex and when it comes to making decisions that can have an impact in the long term. More generally, they have shown that human rationality is much more limited than we thought. Our very ability to reason is extremely reduced and, in many cases, we avoid reasoning, preferring to be guided by our intuitions and our search for immediate satisfactions, even if they are perfectly false for the former and dangerous for the latter. In the analysis and resolution of the ecological crisis, the use of the analytical system is crucially necessary. The intuitive system suffers from many limitations making it unsuitable for solving such a complex problem that requires anticipating what the state of the planet will be in decades or centuries. It is notably the intuitive system and the reasoning biases associated with it that prevent us from correctly apprehending the current environmental situation and adopting the decisions that would be appropriate to deal with it.[18] If for the moment, our handling of the environmental crisis is so catastrophic it is partly because, in general, it is probably more the intuitive system than the analytical system that has presided over our decisions. But all hope is not lost because all the research conducted following Kanheman and Tversky has clearly shown that despite the natural domination of the intuitive system over the analytical system, many factors could contribute to the analytical system taking over the intuitive system. These include the level of

[18] Gifford (2011).

education, the level of scientific and philosophical training and more generally the impact of culture and society on individual psychological functioning.

The astute reader will no doubt have noticed that the neuroscientific, psychoanalytic and psychological analyses have each identified quite similar structural oppositions: power of the reward circuits *vs.* analytical capacity of the neocortex, pleasure principle *vs.* reality principle, intuitive system *vs.* analytical system. The first term of these dichotomies always refers to what is oldest from an evolutionary point of view, probably also the most powerful or at least the least controllable. The second term refers to the most recently acquired psychic abilities in the history of our species: those of critical analysis, reasoning, rationality, the possibility of projecting into the future and planning it, and finally the ability to resist the archaic impulses that animate us all. In this tension between these two components of human nature, the balance can tip one way or the other. There is no doubt that it is up to society, our institutions, culture to tip the balance on the right side. For the moment, the reason why we are where we are, it must be recognised that the first term of these dichotomies has predominated. It is from this imbalance that a large part of the problems we encounter today, including that of the environmental crisis, comes from. The challenge of the centuries to come will be to reverse this hierarchy, if that is possible.

3.7 Conspicuous Consumption: Sex and Power

The sociologist and economist Thorstein Veblen was probably the first to finely analyse the behaviour of consumers of the favoured American class in the nineteenth century (the so-called leisure class) and that of the so-called "nouveaux riches".[19] He is undoubtedly the spiritual father of Baudrillard, Bourdieu, Girard: these subtle analysts of human behaviours intended to affirm social belonging through well-defined codes in on matters of taste, practices and targeted consumption. Veblen focused on describing ostentatious consumption: that which is displayed or exposed, more or less subtly, to communicate one's status or social success to others. He casts a caustic eye on the so-called leisure class, this upper class that reveals its success in proportion to its ability to waste its money through the acquisition of useless luxury items. In short, it is no longer about consuming to satisfy real needs, but rather by virtue of the symbolic value of this or that consumption. Advertisers have understood this well. They rarely seek to arouse the desire for a product by praising its usefulness or technicality; they make sure to trigger symbolic representations that are in line with what the consumer wants to communicate to others or that are in line with the image that a particular consumer good reflects back to them. Undoubtedly, we reach the heights of human stupidity when people, probably in search of an infinitely positive self-image, have come to buy capsules of

[19] Veblen (1899).

24-carat gold leaves costing several hundred dollars which, once ingested, allow them to deposit sparkling and golden excrement, the supreme sign of wealth that can afford the luxury of digesting and thus degrading their own fortune.[20] When others agree to spend tens of thousands of euros to adorn themselves with watches equipped with obsolete technology that does not even compete with modest digital watches worth a dozen euros, it is paying very dearly for a derisory and ridiculous symbolic communication. If the consumer society has played on this need to display one's power, success or heritage, such behaviours are actually very old. In Renaissance Europe, the state rooms of grand houses were incredibly luxurious while the living rooms were of great austerity and often of real discomfort. Even further back, the Potlach ceremony, practised by Native American tribes, followed the ritual of gift-giving. When a chief made a gift to another chief, the latter, to affirm his status and rank, returned the favour by granting a gift of at least equivalent value. The symbolic affirmation of his power and rank was expressed by his ability to part with sometimes precious goods: prestige always requires some sacrifices. In the most extreme cases, among the Kwakiutl of the Northwest Pacific, rival tribal chiefs even came to ostentatiously destroy part of their wealth to signify their power to their rival,[21] a behaviour that is reminiscent of the contemporary propensity of some consumers to waste part of their fortune on luxury items without real aesthetic value and without real utility.

Within the theoretical framework of evolutionary and comparative psychology, it is quite obvious that these ostentatious and culturally coded human behaviours are actually the expression of a sexual competition that all animal species engage in. This obsession with sex and this quest for sexual pleasure results from an evolution that has the immense adaptive advantage of encouraging us to reproduce and thus ensure the transmission of our genes to our descendants: ultimately, therefore, the survival of our species. The power of this drive is infinite. As mentioned above, we are programmed for the pursuit of pleasure and there is hardly any pleasure more intense and fulfilling than sexual pleasure and its romantic sublimation: love. The most fascinating trick that nature has devised over the course of evolution is the orgasm: this intense and immediate pleasure that most seek to obtain whatever the cost. Obviously, if the sexual act was deadly boring, we would not be nearly 8 billion humans on this planet. In nature, especially among birds and fish, males are generally much more brightly coloured than females. They often have large appendages supposed to represent their power (mandibles in insects, hyper-developed fins in fish). When they are not visually prominent, they adopt very visible or very audible behaviours. This is the case, for example, with most songbirds, like the nightingale, which compensates for the sobriety of its plumage with a magnificent song of incredible power and creativity. Large birds of prey, which do not shine either by their plumage or by their song, develop spectacular aerial behaviours or, more prosaically, demonstrate their agility by bringing prey to the female. It is also the males

[20] Hamilton (2010).

[21] Boas (1920).

of most species who engage in dangerous and sometimes deadly fights to assert their power and thus secure a place in the group hierarchy that allows them to have access to the greatest number of females. Among chimpanzees,[22] probably the anthropoid ape closest to us, violence within a group can be extreme to the point of leading to death, and the power thus obtained by the victor always gives the new dominant male the possibility of transmitting his genes to the maximum number of descendants because that is the objective that animates the living. In the most general case, females adopt much more sober and cautious adornments and behaviour because this has the adaptive advantage of going more unnoticed and thus escaping predators, especially during the sensitive period of raising their offspring.

Anthropologists have described perfectly similar behaviours in humans. Whether it is risky behaviour aimed at revealing bravery, behaviours signifying power, agility or strength, protective capacity towards the group, In the hierarchy, in ostentatious behaviours revealing power or wealth, men have always surpassed women. Their rivalry has indeed led them, both in the past and today, to violent conflicts that have continually marked their history. It is enough to analyse the gestures and language of our contemporary statesmen to perceive the violent and impulsive monkey that the codes adopted by humans in their desire to extract themselves from the animality that catches up with them in such a ridiculous way sometimes. As with other animals, the higher the social status, the greater the wealth, the higher the chances of being able to seduce a large number of partners. We must not forget that humans have been polygamous throughout their history and this at least until the appearance of the major religions, but only those who had a significant social status in a hierarchical society really benefited from this practice. The case of great leaders who had hundreds of wives and thousands of children are legion, the most extreme case being that of Genghis Khan who is said to have 16 million descendants on the planet today. The situation has hardly changed today, without of course reaching these orders of magnitude. Statistical studies clearly show that the seductive power of men is partly determined by their social status and their level of wealth.[23] As it is always easier to seduce people of lower social class than the opposite, it is in one's best interest to be as high as possible in the social hierarchy: this offers more choices. From this point of view, the visceral hatred that members of the dominant class have towards the "nouveau riche" is perfectly understandable. The nouveau riche are members of a lower class who, because they have been financially successful, try to appropriate the codes of the upper class in order to finally be able to benefit from its advantages. The ostentation with which they display their newly acquired wealth—large car, jewellery, luxury items…—is a signal often clumsily emitted in order to obtain the status that ensures them the possibility of seducing more individuals. The extreme sensitivity that most parents show regarding the social status of their child's partner is just another expression of the desire to

[22] However, the behaviour of the bonobo differs from that of the chimpanzee (*Pan troglodytes*).

[23] Trump's abominable and vulgar retort is a caricatured expression of this: "When you're a star, you can grab them by the pussy".

preserve their own social status and to prevent sexual competition from involving members of a class normally put out of play. The Indian caste system is never as restrictive as when it comes to choosing a sexual partner.

The economic war waged by states, that waged by companies as well as individuals, are only very sophisticated and sometimes complex expressions of the sexual competition that is at the heart of our psyche. The inequality within our society is another consequence of this competition and the care that the temporary winners take to maintain their status superiority in the never-ending quest for power and the sexual opportunities that result from it. Of course, it is not a question of thinking that we are seeking to rise socially in order to optimise our chances of finding a suitable sexual partner. These are objectives that are not thought about because they escape any form of conscious analysis. These archaic and biologically ancient forces operate without our knowledge and are all the less controllable as they intervene at a level very far from that of our consciousness.[24]

Contrary to the Rousseauist conception of an originally peaceful humanity, our ancestors were already violent. Homicides were infinitely more common than they are today in the most insecure places on the planet[25] and among the most common reasons for aggression between tribes, or homicide within the same tribe, sexual rivalry was high on the list. Ostentatious consumption is ultimately only a very pacified and watered-down form of this fierce desire to achieve dominant status. The relationship between sexuality, social status and competition is very obvious and has been empirically confirmed many times.[26] Everyone knows the role of testosterone, a hormone produced by the gonads in males and in smaller quantities by the ovaries in females. This hormone plays a fundamental role in our sexuality, in our pugnacity and our aggressiveness, as well as contributing to our psychological balance. Psychologists have conducted an experiment in which subjects were randomly selected to drive either a powerful luxury car (a Ferrari for example), or an old and modest car. The analysis of the testosterone level once the driving phase was over, revealed a much higher level of this hormone in the former than in the latter. In a way, driving a vehicle symbolising power and authority had contributed to making their driver more inclined to sexual activity.[27] If we complement this result with the fact that women are on average more attracted to drivers of beautiful cars than to modest old bangers,[28] we have here a remarkable illustration of the relationship between power, status and sexuality. In a similar way, following presidential elections in the USA, it was observed that in the states that had massively supported the winning candidate, the level of consultation of pornographic sites was much higher than in states that supported the losing candidate.[29] The same phenomenon is

[24] Wilson et al. (2012); Bowles et al. (2003).

[25] Gat (2015).

[26] Choi and Bowles (2007); Majolo (2019).

[27] Saad and Vongas (2009).

[28] Dunn and Searle (2010).

[29] Markey and Markey (2011).

observed among supporters of a team, or an individual, who wins a competition. Identifying with the winner is enough to appropriate a part of their success with the positive effect that results in the production of testosterone and the increase in our sexual urges. Ostentatious consumption may seem ridiculous and irrational. However, it is part of a very common form of competition for power, status and therefore sex.

3.8 When Intraspecific Competition Compromises the Survival of the Species

Every living being is driven by two major objectives: to stay alive and to pass on life. In fact, animals, like us, are programmed to avoid death and to give life; the first objective being a necessary condition for the realization of the second. Both of these objectives are subject to different constraints and it is the role of evolutionary processes to ensure that the individual is best adapted to achieve them simultaneously because the survival of an individual who would not leave offspring would lead to the disappearance of the species. Intraspecific competition takes place between individuals of the same species and leads to the fact that the individuals best equipped in this competition will also be those who have the most chances of reproducing and therefore of having offspring. In fact, intraspecific competition is essentially a sexual competition. This latter plays an obvious positive role for the species because the preferences of partners, especially those of the female in most animal species, contribute to favouring the males most capable of achieving reproductive success. Laboratory experiments have further shown that after a few generations, the offspring of two groups of individuals subjected or not subjected to sexual competition—the individuals subjected to sexual competition simply had the opportunity to choose their partner—varied significantly. Sexual competition allowed the production of genetically healthier individuals and more capable of surviving,[30] which led these authors to state that "sexual competition protects from extinction".

However, there is sometimes conflict between intraspecific competition and interspecific competition. In general, evolutionary processes allow an optimal balance to be achieved between these two forms of pressure. To take a classic example, the exceptional plumage of the male peacock is a pure product of sexual selection. It has no other functions than to attract and seduce females. It is a form of ostentatious sexual dimorphism, a beautiful biological form of ostentatious consumption. The more an individual has a remarkable tail, the more spectacular his display, the greater the probability that he will reproduce: this probability is precisely determined by the number of eyespots on the tail. The problem is that such spectacular plumage and a tail of such large dimension can be a real handicap with regard to interspecific competition and therefore the survival of the individual in a natural

[30] Lumley et al. (2015).

environment. Not only could the male thus adorned be handicapped in his movements to escape potential predators, but he will be more easily detectable, which is perhaps why the peacock is now more common in captivity than in its natural environment: it has attracted the attention of humans perhaps even more than that of female peacocks. This is probably what led Charles Darwin in 1860 to say: "The sight of a feather in a Peacock's tail makes me sick", Darwin actually questioning the adaptive interest for the survival of the species of such an appendage. The case of the breast in women is similar. Unlike female chimpanzees or gorillas, women have a disproportionately large breast whose function is not limited to breastfeeding. It is a sexual attribute that has been selected throughout our evolution, not because of its function, but because of the attraction it aroused in men. At the extreme, a large breast, although fascinating for some men, can become a real handicap if it comes to running to escape a predator. This phenomenon is called Fisherian runaway named after the biologist who identified it. It can be summarized by the following formula: when the struggle for sex leads to the endangerment of the species.

On a more global level, this is exactly what is happening to our species. The frenzied competition in which we are engaged leads not only to terrifying violence, especially when organized by states, but even in the absence of such violence, it can be considered that the incredible level of inequality reached in our hierarchical societies, an inevitable consequence of exacerbated competition, generates unbearable suffering for those who are its victims. More directly related to the heart of my argument, it is of course this so powerful and omnipresent sexual competition that fuels our compulsive desire to consume because it is through our consumption that we best express our status and our power: our value as a potential reproducer. For this, we must selfishly enrich ourselves and thus contribute to the economic success of our societies: the famous invisible hand of Adam Smith is just one of the remarkable consequences of the archaic biological forces that drive us. To take a well-known image, the man at the wheel of his shiny Jaguar is not much different from the nightingale that exhausts itself singing or the peacock that displays its tail, the problem being that, given our ability to produce infinitely and to consume without it corresponding to a real need, we are not in competition with other species. Predatory animals—as might be the case with the peacock—but in conflict with a planetary ecosystem that cannot survive and maintain itself in the face of our insatiability. It is from the confrontation between our infinite desire and the finitude of our planet that all contemporary environmental problems arise. The whole question is thus whether the human species can escape the biological forces that have effectively propelled it to the forefront of the stage in a few tens of thousands of years and which could now put an end to its exceptional adventure. What was adaptive in the past is clearly no longer so today. Being aware of this is one thing, mastering it is another.

3.9 When Society Organises Itself Around the Archaic Biological Forces that Drive Us

If the evolution of human is, as for all animal species, biologically constrained, it is, more than for any other species, largely culturally determined. This is the phenomenon of gene-culture coevolution: genes determine our culture and culture contributes to our genetic evolution. Culture evolves at a much faster pace than genes, creating new environments that exert new selective pressure on genes. For example, we know that the cultural upheavals caused by the advent of agriculture 10,000 years ago led to a massive wave of genetic modifications.[31] It is likely that this phenomenon is accelerating even today given the frenetic pace of cultural evolution. Therefore, our culture can contribute to the emergence of certain behaviours or certain physiological as well as morphological traits. It can do so all the more because the cultural system allows for complex adaptations that are passed on from generation to generation in a relatively independent manner with respect to biological evolution. This, of course, explains the incredible cultural diversity of human societies. Therefore, the question that arises is whether the evolution of our culture, especially its current characteristics, is likely to counteract the biological forces that underlie our consumption or, on the contrary, to resonate with it in a powerful and dangerous alliance of the biological and the cultural.

Today, without a doubt, it is this second hypothesis that must be favoured. To understand this, it is necessary to recall certain historical and empirical elements of what is commonly called consumer society. While humans have been increasing their consumption for 10,000 years, the movement has been very slow and almost imperceptible until the time of the colonisation of the planet by European countries. Then it gradually accelerated from the industrial revolution and followed an exponential curve since the 1950s. It was in the 1930s, first in the USA, then in almost all of the planet, that the state, businesses and trade unions organised society in such a way as to stimulate consumption, which has become one of the main engines of our economies and an essential component of human life. The results have been extraordinary. The production of goods and services related to domestic consumption doubled from 1946 to 1956, then again from 1956 to 1970, to the point of representing two-thirds of the gross national product in the USA. In reality, the organisation of mass consumption appeared as a necessity in the face of the increasing capacities of a production apparatus that had become hyper-powerful. The dangers of overproduction and unemployment could only be resolved by a constant increase in consumption. We have in a few years moved from an "ant" culture to a "cicada" culture. If the frugal and economical behaviour of our ancestors was still valued at the beginning of the twentieth century, it was then stigmatised in the "American way of life" which considers it as a mean stinginess, a symbol of a backward peasantry that has not grasped the so enjoyable, modern and libertarian character of this new relationship to the world.

[31] Henrich and Boyd (2002).

It is quite remarkable that the pioneer of scientifically organised political and advertising propaganda, Edward Bernays, was Freud's nephew. He had, no doubt influenced by his uncle's work, perfectly understood the importance of the symbolic, phantasmatic, libidinal and hedonistic dimension of our consumer behaviour. Playing and manipulating it appeared as the most appropriate way to methodically and effectively guide consumer behaviour. He is notably responsible for the invention of the traditional American breakfast as well as the consumption of tobacco by women, then perceived as an expression of a form of female liberation. The "Employment Act" of 1946 in the USA is the institutional expression of this obsession with consumption, which required federal governments to promote employment, production and purchasing power: the three virtuous and inseparable elements of a healthy economy, but also of a happy democracy, confident in itself and placing individual well-being and success at the heart of its objectives. It was from this time that credit developed to an extreme degree. Borrowing was no longer seen as a dangerous and criticisable decision, but as the most beautiful manifestation of a modern, optimistic and hedonistic conception of life, freed from the restraint and control so dear to the ancients. The result of easy credit, notably the aptly named consumer credit, was incredible. American household debt exploded, increasing by 10 billion per month in the 1980s, then by 25 billion per month in the mid-2000s. As a proportion of annual income, this debt was in the order of 60% in the 1950s, reaching 100% in the 1980s and then 130% in 2005.[32] In December 2019, the debt of American households reached $14,159 billion after 22 consecutive quarters of growth. The political and economic objective has been achieved. By arousing desire, by valuing actions likely to satisfy it immediately, the whole society has embarked on a mad race, often catastrophic for the households themselves, very economically dangerous in the medium term and absolutely terrifying in the long term at the ecological level.

Arousing desire has become a major axis of any form of economic and industrial project. It has taken place in a society where vast cultural changes resulting from the social movements of the 1960s and 1970s have opened the era of individualisation. Previously, each individual took place in a fairly homogeneous society and was locked in a complex system of norms and values: tradition left little room for singular behaviours and sometimes reprimanded them. Now, each individual has the freedom to create his own identity. He almost has the duty. He is no longer or no longer believes he is the product determined by his birth and his social class of belonging. He can, through a "chosen" consumption and only constrained by his financial means, express what he is or what he would like to be. Consuming is expressing one's identity and singularity in a project in perpetual construction. Undoubtedly, this has exerted a force all the greater as with the era of assembly line work and a parcelled work devoid of meanings, the employee could no longer find in his professional activity an authentic space for individual expression. Only consumption could now play this function by allowing the consumer to transmit to others an

[32] Rampell (2009).

idealised image of himself through a consumption that has become eminently communicative. It was then up to advertising to play on this now crucial and absolutely fundamental symbolic dimension to arouse the desire to consume. Advertising then became so important that its cost sometimes became higher than the cost of the object it was promoting. And as minds are never as malleable as when they are young, advertising investment has been exceptional towards children and teenagers, willing victims of the dictatorship of brands. In 1983, American companies spent $100 million on advertising specifically addressed to children. In 2000, this investment reached more than $17 billion. It is estimated that each American child has seen on television, or now on the Internet, several tens of thousands of advertisements from birth to adolescence.[33]

What is absolutely fascinating here is that advertisers have very precisely exploited the biological and psychological determinisms described above as being at work in our addiction to consumption. Pleasure above all and immediate pleasure, sex and the search for social status are present in almost all advertising content. Almost none of the current advertisements escape one or the other of these components. Sex is almost systematically used implicitly or explicitly in advertising messages regardless of the type of culture that has been studied.[34] Social status holds a considerable place and is often associated with clearly sexual or eroticised messages. We will simply note in recent years a trend towards greening these messages because affirming one's interest in defending the environment has become a convenient way to absolve the consumer about to succumb to obviously anti-ecological purchases, such as this recent advertisement for a car brand boasting the hybrid technology used by powerful and polluting SUVs.

In the end, this consumer society, and the great surface sophistication of the advertising messages associated with it, only exploit what is most emotional and biologically archaic in Man. Without having a clear and explicit understanding of it, advertisers have been able to take advantage of the strength of the hedonic system, the power of the pleasure principle and the domination of the intuitive system over the analytical system. The overall effectiveness of the system leading us to consume is formidable because it is very difficult for a human to resist forces so constitutive of our nature. All the more difficult as advertising messages precisely aim to solicit the psychological component most sensitive to the satisfaction of immediate pleasure and capable of inhibiting the cognitive component likely to resist it. And as desire is by nature infinite and can be frequently renewed due to the frequency of our technological innovations, our economies can count on our infinite appetite for additional consumption.

Where the process becomes perverse is when the carefully instilled frustration leads the avid consumer to get into financial difficulty to satisfy constantly renewed desires without their satisfaction being able to bring the individual the sense or tranquillity likely to free him from his dependence. Financial difficulties affect an

[33] Hamilton (2010).

[34] Belk et al. (2003).

increasing number of individuals, including the upper middle class, and require raising one's salary level and therefore working more. The individual is thus put under permanent tension and is confronted with a state of psychological exhaustion today very commonplace, a state of psychological exhaustion further exacerbated by the professional and wage precariousness now the norm within our liberal economies. However, this state of psychological exhaustion is at the origin of a phenomenon with formidable consequences: it is the so-called ego depletion phenomenon. Psychologists[35] have thus shown empirically that as soon as individuals were subjected to even modest cognitive effort, this had two joint effects of great importance. The first is the increase in the difficulty encountered by individuals in delaying the satisfaction of a desire, even when the latter is identified as problematic or dangerous. In other words, ego depletion leads to a reduced ability to exercise willpower and control impulses. It makes any form of addiction more powerful. The second effect is to inhibit the analytical system and favour the intuitive system.[36] In a way, ego depletion reduces the capacity for reasoning and rational analysis and weakens the barriers that individuals might build to protect themselves from behaviours that are harmful to them. In short, we are not only subjected to the clever and incessant solicitations of advertising, but the very organisation of society contributes to a weakening of our ability to resist the natural desire for consumption. All this leaves little possibility of escaping from this infernal loop which is characteristic of all addictive behaviour. This is indeed the whole problem: our society is above all a society of addiction or which has made addiction one of the basic engines of its economy.

Just as the addictive person still aware of their dependence passes a severe judgement on the drug they are victim to, most Western consumers generally take a rather critical view of this almost pathological form of materialism. Cross-cultural studies have shown that excessive consumerism and our dependence on the acquisition or enjoyment of material goods are negatively evaluated.[37] Already, Hobbes, in his time, had noticed the chronic anxiety of a bourgeois society obsessed with material values. The philosopher and sociologist Herbert Marcuse, heavily influenced by Marx and Freud, had deplored our dependence on what he called false needs, that is to say, needs created from scratch and which, once satisfied, do not really increase our well-being. But if such awareness has undoubtedly emerged over the past 50 years, it is not yet capable of curbing our addiction because, for many consumers, bad consumption is generally that of others, rarely their own. The question therefore remains as to whether another form of societal organisation would be capable of curbing our compulsion to consume, now perfectly integrated into the lives of all individuals, including the youngest. Without this, there will be no solution to the ecological crisis, the myth of green consumption being there only to allow us to consume in good conscience and to feed an economy that only survives thanks to our consenting servility as docile consumers.

---

[35] Baumeister et al. (2008).

[36] Kelemen and Rosset (2009).

[37] Ger and Belk (1996).

References

Baumeister, R. F., Sparks, E. A., Stillman, T. F., & Vohs, K. D. (2008). Free will in consumer behaviour: Self-control, ego depletion, and choice. *Journal of Consumer Psychology, 18*, 4–13.

Belk, R. W., Lizger, G., & Askegaard, S. (2003). The fire of desire: A multisited inquiry into consumer passion. *Journal of Consumer Research, 30*, 326–351.

Biermeyer, P. (2001). *Coming changes in the U.S. clothes washer market*. Lawrence Berkeley National Laboratory. Unpublished paper.

Boas, F. (1920). The social organisation of Kwakiutl. *American Anthropologist, 22*, 111–126.

Bowles, S., Choi, J.-K., & Hopfensitz, A. (2003). The coevolution of individual behaviours and group level institutions. *Journal of Theoretical Biology, 223*(2), 135–147.

Choi, J.-K., & Bowles, S. (2007). The coevolution of parochial altruism and war. *Science, 318*(26), 636–640.

Crooks, K. R., & Soule, M. E. (1999). Mesopredator release and avifaunal extinctions in a fragmented system. *Nature, 400*, 563–566.

De La Bruslerie, H. (2012). The psychological value of time: A synthesis of the literature. *Economic News, 88*(3), 361–400.

Dunn, M. J., & Searle, R. (2010). Effect of manipulated prestige-car ownership on both sex attractiveness ratings. *British Journal of Psychology, 101*(Pt 1), 69–80.

Fear, J. (2008). *Stuff happens: Unused things cluttering up our homes.*, Research Paper No. 52. The Australia Institute.

Gat, A. (2015). Proving communal warfare among hunter-gatherers: The quasi-rousseauan error. *Evolutionary Anthropology, 24*, 111–126.

Ger, G. L., & Belk, R. (1996). I'd like to buy the world a coke: Consumptionscapes of the 'less affluent world'. *Journal of Consumer Policy, 19*(3), 1–34.

Gifford, R. (2011). The dragons of inaction psychological barriers that limit climate change mitigation and adaptation. *American Psychological Association, 66*(4), 290–302.

Hamilton, C. (2010). *Requiem for a species*. Earthscan.

Harris, C., & Laibson, D. (2001). Dynamic choices of hyperbolic consumers. *Econometrica, 69*(4), 935–957.

Heather, N., & Segal, G. (2013). Understanding addiction: Donald Davidson and the problem of akrasia. *Addiction Research and Theory, 21*, 445–452.

Henden, E. (2012). Addictive actions. *Philosophical Psychology, 26*, 1–21. https://doi.org/10.1080/09515089.2012.660688

Henden, E., Melberg, H. O., & Røgeberg, O. J. (2013). Addiction: Choice or compulsion? *Frontiers in Psychiatry, 4*(77), 1–11.

Henrich, J., & Boyd, R. (2002). On modelling cognition and culture: Why cultural evolution does not require replication of representations. *Journal of Cognition and Culture, 2*, 87–112.

Jackson, T., & Marks, N. (1999). Consumption, sustainable well-being and human needs—With reference to UK expenditure patterns 1954–1994. *Ecological Economics, 28*(3), 421–442.

Kahneman, D. (2011). *Thinking, fast and slow*. Penguin.

Kelemen, D., & Rosset, R. (2009). The human function compunction: Teleological explanation in adults. *Cognition, 111*, 138–143.

Loss, S. R., Will, T., & Marra, P. (2012). Direct human-caused mortality of birds: Improving quantification of magnitude and assessment of population impacts. *Frontiers in Ecology and the Environment, 10*, 357–356.

Lumley, A., Michalczyk, L., & Kitson, J. (2015). Sexual selection protects against extinction. *Nature, 522*, 470–473.

Majolo, B. (2019). Warfare in an evolutionary perspective. *Evolutionary Anthropology, 28*, 321–331.

Markey, P., & Markey, C. (2011). Pornography-seeking behaviours following midterm political elections in the United States: A replication of the challenge hypothesis. *Computers in Human Behaviour, 27*, 1262–1264.

Okin, G. S. (2017). Environmental impacts of food consumption by dogs and cats. *PLoS One, 12*, e0181301. https://doi.org/10.1371/journal.pone

Pronin, E. (2007). Perception and misperception of bias in human judgment. *Trends in Cognitive Sciences, 11*(1), 37–43.

Rampell, C. (2009). Shift from spending to saving may be slump's lasting impact. *New York Times*, 10 May 2009.

Robson, A. J. (2002). Evolution of human nature. *Journal of Economic Perspectives-Volume, 16*(2), 89–106.

Ropke, I. (1999). Theories of practice—New inspiration for ecological economic studies on consumption. *Ecological Economics, 68*, 2490–2497.

Saad, G., & Vongas, J. G. (2009). The effect of conspicuous consumption on men's testosterone levels. *Organizational Behavior and Human Decision Processes, 110*, 80–92.

Samuelson, R. (2005). Homes as hummers. *Washington Post*, 13 July 2005.

Shove, E. (2003). *Comfort, cleanliness and convenience. The social organisation of normality*. Berg.

Thaler, R. (1981). Some empirical evidence on dynamic inconsistency. *Economics Letters, 8*, 201–207.

Veblen, T. (1899). *The theory of the leisure class*. (1998 edition) (Great minds series). Prometheus Books.

Warner, J., & Pleeter, S. (2001). The personal discount rate: Evidence from military downsizing programs. *American Economic Review, 91*, 33–53.

Wilson, M. L., Kahlenberg, S. M., Wells, M., & Wrangham, R. W. (2012). Ecological and social factors affect the occurrence and outcomes of intergroup encounters in chimpanzees. *Animal Behaviour, 83*, 277–291.

Chapter 4
The Economy as a Destructive Force of the Planet

Abstract There is a perfect alignment, a complex form of integration and synergy between our economic objectives (at the individual, corporate, or state level) and our addiction to consumption. Every individual and human group is caught between our propensity to consume and the need for economic growth. The major difficulty faced by individuals and states is that economic competition makes any form of consumption reduction impossible. Economic growth is, in the short term, necessary for states, while it is lethal for humanity in the long term. It is therefore clear that the fundamental constraint weighing on each national economy is incompatible with the preservation of our planet.

From a certain point of view, the economy could very well be integrated into ecology as a branch of life sciences because it ultimately describes the exchanges and energy flows that take place between humans and between humans and nature. Society, and the economy that sustains it, can be identified as a macro-organism characterised by a giant metabolism, drawing on the planet's resources, producing wealth and waste that we must recycle or disperse in the environment. It would be naive to imagine that the economy is decoupled from general biological and ecological constraints: it is extremely dependent on them. It is also likely that economists would take a very different view of the world if they crossed economic science with biological sciences more often. The few economists trained in biology are all heterodox because taking into account the biophysical constraints of the Earth has allowed them to have a different view of the fetishism of growth. Unless one is demonstrating a form of dramatic utopia or a serious form of denial, the current economic operation is simply not compatible with even a very minimalist preservation of the planet. The economy is indeed at the heart of the ecological problem that concerns us. It is also certainly its solution even if no one, to date, is really and practically capable of tracing even vague outlines of it. The complexity of the task

T. Ripoll, *Human Psychology and the Destruction of the Planet*, Anthropocene – Humanities and Social Sciences, https://doi.org/10.1007/978-3-032-02080-2_4

lies in the fact that the functioning of our economies is subject to multiple constraints that make up a system[1] and it is very difficult to evolve this system without generating formidable upheavals: these constraints are ecological, psychological and political in nature. Exploring the nature of these constraints and their intertwinement will be the objective of this chapter.

4.1 Growth or Death; Growth and Death

It is not so much the morbid character of this title that should attract attention as its aporetic dimension. It is in this latter that lies the essence of the impasse in which we find ourselves. Without growth, our economies die, causing in their fall, formidable social and political disturbances. Keynes himself had clearly identified it: "The main objective of the economy is to ensure social stability by producing the wealth capable of increasing our material well-being and freeing us from natural hazards". The current Chinese government does not even hide it and officially displays its objective: to buy social peace and avoid political contestation by increasing the standard of living of as many people as possible. Cynicism or realism, one must give the people something to consume if one wants to control them. It can also be thought that if the Soviet bloc collapsed, it was more because of the envious gaze that the citizens of these countries had on the opulence and growth from Western countries due to the restriction of freedom they were then suffering. If populism is flourishing in Europe as well as in the USA since the 2008 crisis, it is partly due to the period of recession or stagnation that this crisis has caused.[2] Periods of recession are the breeding ground on which populisms, nationalisms and political authoritarianisms develop, all playing on their rejection of the system while they are its most tragic expression.

Clearly, the establishment of modern democracies has gone hand in hand with an increase in living standards and it would be a spoilt child's behaviour to deny what the growth of the last century has brought us in terms of comfort, stability, quality of life, intellectual, cultural development and democracy. But pursuing our growth leads us directly to major ecological problems which, in turn, will constitute even more prohibitive obstacles to the development of our economies, our democracies and life itself. Faced with this double constraint, humans have made a choice: that of growth, that of the economy. For the first time in human history, the economy is an end in itself. We no longer produce and consume to satisfy real needs or to make the world more pleasant. We produce and consume to ensure the minimum growth without which it would be economic, cultural and political chaos: the collapse of our societies. A few weeks after the end of the lockdown imposed by the coronavirus crisis, an economic journalist from France Télévision clearly expressed the nature of the problem caused by this crisis and its possible resolution. The French, she noted, saved 55 billion during the lockdown because they could no longer consume freely;

[1] Gowdy (2020).

[2] Büchs and Koch (2019).

we must now make them want to spend this money, otherwise the risk is great of starting a recession phase that could be fatal to us. The mass is thus said. It is not about consuming to satisfy this or that need. It is about consuming to make the productive machine work, avoid overproduction and ensure the minimum level of growth (about 2%) that can allow the repayment of the debt and the reduction of the unemployment rate. We are engaged in a mad, circular and absurd race: for the economy to survive, for it to continue to produce ever more wealth, citizens must return to their compulsive consumption even if it brings them no gain in terms of well-being and even if the ecological cost of such operation is devastating. The most caricatural reasoning on this subject was clearly expressed by an American magnate of hydraulic fracturing oil extraction who observed that with the decrease in road traffic, traffic jams became rarer, too rare. Because traffic jams lead to an overconsumption of petrol, their rarity causes great damage to such companies, notably an accumulation of problematic stocks and a halt in production leading to the unemployment of many of their employees. Regretting traffic jams is the most pathetic illustration of a superficially rational system, but in reality deeply absurd. It is the same type of absurdity that leads to the fact that the sinking of an oil tanker, and the resulting oil spill, contribute to the increase in GDP: the ultimate goal of any government. A stable state of the economy, without even going as far as recession or depression, would lead to its collapse, which is why everything must be done to ensure that we follow a continuous growth curve. Our governments are thus faced with an almost impossible mission since they must meet two contradictory objectives: ensure the economic growth necessary for social stability and work to protect our environment.[3] This contradiction is only a pale reflection of those that run through the whole of the population, itself in constant search for an increase in income or wealth while demanding drastic measures to protect the planet. As the economic constraint intervenes in the short term and the ecological constraint in the long term, and as we are psychologically more constrained by the short-term than by the medium or long-term, it is today, and probably still doubt tomorrow, the economic constraint that will dominate our societies. The slogan of Nicolas Hulot, quickly taken up by the yellow vests movement "end of the world and end of the month, same fight" expresses the bet, perhaps impossible, to satisfy these two constraints simultaneously.

4.2 Growth, Expansionism and Inequalities: A Triptych Doomed to Collapse

Economic growth seems indispensable for multiple reasons, but all ultimately result from an undeniable technical reality. Human productivity has continued to increase and it has done so exponentially for two centuries. Each modern human is capable of producing what previously tens, hundreds, or even thousands of humans produced in certain sectors of activity. The consequence is of great banality: if a human

[3] Costanza et al. (2016); Kallis et al. (2018).

is capable of producing what tens of others did in the past and if our consumption remains stable then millions of humans can do without working. In reality, millions of humans will be forced into unemployment—generally the least educated—with all the disastrous consequences, psychologically and economically speaking, that result from it. The general equation of this phenomenon is of great simplicity. If the growth of GDP is equivalent to the increase in productivity, employment is stable. If GDP growth is lower than productivity increase, the employment rate decreases and if it is higher than productivity increase, the employment rate increases. In other words, given the natural increase in productivity, humans must always consume more in order to increase GDP, with the annual GDP growth corresponding to a country's growth rate. There is a clear convergence here between the psychological insatiability of each individual and the growth requirement inherent in the functioning of our economic and political system. This is indeed a convergence, not a coincidence, our unsatisfied desire for consumption being one of the drivers of our economic functioning and the latter relying on this fundamental insatiability. We are thus many to complain about an economy that we disapprove of in many respects and yet we feed it more and more through consumption of which we are no longer even aware and which continues to grow every day.

What are the factors that have contributed and still contribute to the increase in human productivity? They are essentially of two kinds. The first concerns the discovery and exploitation of fossil energy (oil, gas, coal). The industrial revolution would not have taken place and our contemporary societies would hardly have differed from those that preceded the industrial revolution without this energy. All current technology, all our scientific developments, all our culture, everything that constitutes our environment and which seems so ordinary to us today, would not exist without this bounty that nature has bequeathed to us. Our economies are not off the ground. They have been and will always be dependent on the energy we can extract from the planet. Jean-Marc Jancovici,[4] provides a very telling analysis of the importance of energy aspects in our society. It is easy to measure the amount of energy consumed by a human for his own metabolism and the energy he can return for all sorts of basic mechanical activity: moving, moving objects, working the land... Roughly speaking, a human doing intense physical work over a day consumes 5 kwh. But the mechanical energy he will be able to return from this consumption will be much more low, of the order of 0.05–0.5 kwh depending on the type of physical activity. The energy efficiency of a human is modest and the few technological innovations that preceded the industrial revolution have hardly changed the situation. The power leap achieved thanks to the transformation of fossil energy is evident if we recall that a litre of petrol contains 10 kwh and that the efficiency of a modest thermal engine is of the order of 30%. In other words, with 1 litre of petrol, you can recover 3 kwh, i.e. the mechanical energy that would result from the work of several dozen humans. It is possible today to quantify the amount

[4] See the website https://jancovici.com/transition-energetique/l-energie-et-nous/combien-suis-je-un-esclavagiste/ for a very accessible and didactic explanation by this author.

of energy consumed by each human. It was estimated in 2012[5] that each French consumed 30,000 kwh: much less than an average American and much more than a Somali. Based on the work produced by a human who only has his own physical strength, it is possible to calculate the number of "energy slaves"[6] necessary for maintaining the lifestyle of an average French person: the number of humans who would be necessary to maintain our lifestyle if we did not have fossil energy. For agriculture, 23 energy slaves would be necessary, 145 for industry, 22 for transport and 237 for residential and tertiary. In other words, without fossil energy, each French person would need about 500 slaves daily to maintain his standard of living. For an American, 880 slaves would be necessary, 2435 for a Qatari and 80 for an Indian, less than 10 for a Nigerian. These astronomical figures, even if they are approximate, reveal both the level of inequality we have reached and how much our extreme level of wealth in the west is largely due to the fossil energy we have: even a French minimum wage earner is probably richer, in terms of energy consumed, than the wealthiest men of the middle ages. These figures reveal above all that productivity has naturally increased at least in the same proportion since fossil energy can effectively replace the work that men had to produce before. For example, for agriculture, if we did not have energy slaves, it would take several hundred million farmers to feed 67 million French people when today there are only 600,000. The second reason for the increase in productivity results directly from the rationalisation and organisation of work on the one hand, from technological innovation on the other hand. The digital revolution now introduced in almost all sectors of activity is in the process of increasing our productivity in an extraordinary way. Artificial intelligence can directly replace this time cognitive work, and no longer exclusively physical work, that only humans were capable of providing a few years ago.

The conjunction of very cheap fossil energy and technological innovation will directly sideline hundreds of millions of humans if GDP does not grow as fast as productivity increases, which is why we are caught up in an endless movement of continuous expansion. By nature, capitalism is expansionist because if expansion is not possible, collapse threatens. This expansionist requirement is so strong that it is considered by economists and historians as a condition for the survival of our economies and empires. Colonisation has of course participated in this dynamic by suddenly offering us almost free goods that we have plundered from indigenous peoples and by opening up new markets that have Incredibly enriched, that of the slave trade and that of the flow of our manufactured products for example. Capitalism demands conquest, whether it be territorial, economic or political, to ensure its survival. However, the system is not very resilient because the mechanisms that lead to growth and expansion also lead to collapse as soon as the growth dynamic is stopped by whatever cause it may be, which is why, economists and governments fear more than anything the disastrous and uncontrollable dynamic that could result from the onset of recession. The logic of credit is a counter to this threat because it imposes

[5] Figures from the General Commission for Sustainable Development.

[6] The concept of energy slave was originally developed by Richard Buckminster Fuller.

and makes possible a forward march alone capable of perpetuating a system by nature unstable since it is incapable of reaching a steady state. But it also contributes to an absurd expansionist dynamic if we keep in the background the awareness of being an inhabitant of a finite and closed space. By injecting billions into the economy, as was the case after the 2008 crisis or the coronavirus crisis, the economy gives itself a little oxygen temporarily while creating a debt that only growth and therefore consumption will allow to mop up, with their usual and inevitable lot of environmental destruction.[7] The same paradoxical solution is found on the issue of pension funding. All governments try to stimulate demographic growth because it is the working population that allows pensions to be financed. The age pyramid cannot be reversed at the risk of a collapse of the pension system. But as today's workers will be tomorrow's retirees, the movement must always be increasing: having children to pay for future pensions, that's what we are bound to. But until when? This economic logic, perfectly rational in the short term, is perfectly absurd in the long term. In the end, we have created a society and an associated economy that can only fall in the more or less long term. The image that best fits this dynamic is that of a cyclist who must follow a track that leads to a precipice. If the cyclist stops, he falls because he doesn't know how to stay in place. To avoid falling—this would correspond to the classic economic crisis—he strives to move forward with more or less awareness of what awaits him at the end of the track. Because, ahead, the danger is much greater: the precipice, that is radical collapse. As our psyche is always more determined by the short term than the long term, the cyclist will continue to progress and it is only when the precipice gets really close that he may consent to either change direction—a radical societal change still to be imagined—, or to accept the fall linked to his inability to stay in place—the economic crisis inherent in any form of stable state or even worse, of decline. Unlike previous generations, we are aware of the precipice. But it is still too far away to provoke a major societal rupture especially as citizens of rich countries, we enjoy an opulence and quality of life that would have even been coveted by the richest sovereigns of our past: consciousness and rationality do not weigh heavily against hedonism and other psychological forces that drive us towards the precipice.

The expansionism of our capitalist societies is only the product emerging from a form of expansionism already present at the individual level. There is a form of homothety here since the expansionism of states directly echoes, on a larger scale of course, the desire we individually have to enrich ourselves in order to maximise our chances of staying alive and passing it on. Anthropologists have thus shown that in natural disasters or tribal conflicts, it is always the richest families who have survived. The most extreme case is that of the Micronesians of the Pingelap atoll. After the destruction of this atoll caused by a cyclone in 1775, out of 1000 inhabitants, only 20 survived: they were exclusively hereditary chiefs and their close family. Individual wealth protects from disease—this is still the case today—and keeps

[7] President Biden's recent stimulus plan, amounting to 2000 billion, has been praised by many economists and its effects have been immediate. Despite the planned investment in renewable energies, there is no doubt that its ecological balance will be disastrous.

death at bay. Expansionism is Darwinian in the sense that each species tries to exploit its environment as best it can in order to increase its population, its geographical distribution and generally continue to grow, growth being the most direct and effective way to guarantee the survival of the species. Breaking with expansionism as breaking with growth requires us to free ourselves from biological determinisms that our society has so far exploited wonderfully to build our civilisation, both brilliant and despairing. But this expansionism has, in addition to the ecological disaster that results from it today, social and geostrategic implications just as formidable. If it manifests itself at the individual level, it also manifests itself at the level of states and at the level of companies. In a finite universe, with finite resources, the expansion of some can only be done at the expense of others. This is why states and companies are engaged in fierce competition: never as much as during these last decades, with the unchallenged domination of economic liberalism, has the frequency of use of the term competitiveness been so important. For liberal economists, inspired by a brutal form of Darwinism, the game of free competition and a competition as unregulated as possible by states, is the best way to win the bet of growth because it is first and foremost up to companies to create the wealth that states need. The idea, once classic, that governments should protect citizens from the excesses of free enterprise has been replaced by the idea that governments should protect the free market and free enterprise from the excesses of state regulation. The collapse of socialist economies seems to give credence to this dismal conception of our society. This is what makes the liberal economist Milton Friedman, founder of the Chicago school and a major inspiration for the policies pursued by Reagan, Thatcher and Pinochet that free competition and the fundamental inequality that results from it are absolutely necessary to ensure strong growth. That the least adapted companies disappear and the most efficient survive would only be the economic avatar of the laws that govern the natural world. Such a conception is just a radicalisation of Adam Smith's thought that capitalism obeys a natural order and is established from the natural propensity of men to selfishness, to exchange and to the search for the dominant position. Inequality between states, between companies and between individuals is both a normal consequence of a mode of natural functioning of the economy and the very engine of the economy. Inequality is the spur that incites free individuals to make sure to enrich themselves and, in trying to enrich themselves, to contribute, almost without their knowing it, to national expansion and growth. This is how deregulation has become a rule, a principle to which states themselves, including unfortunately within the European community, can only very difficultly derogate. And because wealth is produced by companies and states are dependent on it, the economy is increasingly dissociated from politics, companies taking power over it and governing the course of the world which is increasingly clearly escaping from the democratic power held by citizens. The generalised competition supposed to be necessary for growth thus has three major disadvantages.

The first is to maintain or increase a level of inequality between individuals that ends up appearing normal to us as we have become accustomed to it and which is nevertheless unacceptable and dangerous. The Gini coefficient, measuring the level of inequality within a population, has indeed continued to increase at the level of the planet although it has remained stable in France, which is an exception. Let us recall that at the level of the planet, 80% of the world's wealth is held by 1% of the population.[8] It is important to understand that inequality is not a contingent dimension of liberal economy. It is systemic in nature in the sense that it is necessarily part of the characteristics of the system.

The second disadvantage is that competition between states gives rise in the best case to trade war and in the worst case to military conflicts. Economic competition is the soil on which armed conflicts directly or indirectly develop that have never ceased to punctuate the history of the planet.

The third disadvantage is ecological. Taking into account the environmental dimension, like the social dimension, generally leads to a decrease in competitiveness which is hardly possible in a context of economic war. Trump understood this in a certain sense since, obsessed with the competitiveness of his national economy, he even made the unilateral decision to break with the Paris climate agreements. Let's take a second example, entirely fictional and extreme to illustrate it, but which has the advantage of not stigmatising only the state or companies. Suppose that to preserve the environment, the French car industry is forced by the state to produce only medium-sized vehicles, of low power, without any unnecessary equipment and whose speed could not exceed 110 km/h. This would certainly contribute to significantly reducing the carbon footprint of this type of industry and would allow the marketing of vehicles at unbeatable prices. What would happen then? Immediately, turnover would decrease and a large number of French people, probably among them people sensitive to the environmental cause, would buy more powerful and attractive foreign cars. This would have the dramatic consequence of putting the national car industry into bankruptcy and increasing France's trade deficit. This would have a phenomenal impact on the unemployment rate, one job in ten being directly or indirectly linked to the car industry in our country. Another very real and current example is that of glyphosate. From a technical point of view, glyphosate is a very effective weed killer that allows a significant gain in productivity and therefore greater competitiveness. If it is so difficult to ban its use, it is because such a decision cannot be unilateral. This would penalise too much the farmers affected by such legislation and, in a second step, the states that would have made this decision. The competition that remains at the heart of the functioning of our societies is generally incompatible with the preservation of the environment.

It is now necessary to underline a last point, and not the least, which leaves little hope for our current way of life to continue. Let us recall that our wealth today is directly linked to the presence of a very cheap fossil energy. This energy would disappear tomorrow and our society would collapse almost immediately: nothing that makes up our daily life could resist such a shortage. In fact, there is a close and unquestionable relationship between growth, GDP and the energy available and consumed. This relationship is not very original if we keep in mind that our

[8] According to the non-governmental organisation OXFAM.

societies are organisms with a metabolism like all living beings. The more society grows, the more its metabolism requires energy and the more it consumes. To believe that a physical phenomenon of this type, resulting from the very principles of thermodynamics, could not concern our societies is a utopia or a misunderstanding of the laws of physics. In fact, the main obstacle to our growth will simply lie in the fact that the almost free fossil energy we exploit is on the decline. The peak of production has probably already been reached for oil and it will be quite quickly for gas and coal especially as global energy consumption continues to grow. At the current rate—our energy consumption has increased by 109% since 1973 to 2017—, these non-renewable energies will have disappeared, or will no longer be profitable given their operating cost, in 50 years for oil, 51 years for natural gas, 132 years for coal and 90 for uranium (if we take into account nuclear energy). The concept of Energy Return on Investment (EROI) allows us to better understand the rapid evolution we are facing. The EROI corresponds to the ratio between the energy required for extraction, transport and transformation of any energy material (crude oil for example) and the amount of consumable energy thus obtained (the petrol you put in your car). At the beginning of the oil industry, 1 barrel of oil allowed the extraction and refining of 100 barrels. Never has an energy been so profitable. A real jackpot that was going to enrich, albeit unevenly, the whole planet. The EROI has continued to deteriorate to become lamentably low with the extraction of shale oil. It is currently at best 1 barrel for 10 products in Saudi Arabia. The miraculous solution of renewable energies should not lull us into illusions. The EROI of these energies is relatively low and their exploitation poses formidable technical and ecological problems.[9] In many cases, a complete and honest assessment of renewable energies reveals that their ecological impact remains considerable, including in terms of CO_2 emissions. Their exploitation also consists of externalising our ecological footprint in Africa, China and South America, due to the massive environmental destruction caused by the extraction of minerals necessary for the manufacture of batteries for electric cars for example. While it makes sense to support their development, it would be naive to believe that they can fully replace current fossil fuels and support constant and sufficient growth for our economic functioning.[10] Let's just remember that even if 100% of electricity was produced by renewable energies, it would only account for 18% of our global energy needs, the share of electric energy being only 18% even today. I would not be far from making this quote from economist Kenneth Boulding[11] my own: "Anyone who believes that exponential growth can be ensured indefinitely is either a fool or an economist". One thing is clear. It is mathematically, physically and biologically impossible to have exponential, or simply constant, energy consumption growth in a finite world. Regardless of our *hubris*, regardless of our excessive pretension to believe ourselves capable of freeing ourselves from biophysical constraints, we will very quickly have to face energy shortage. It is very

[9] Moriarty and Honnery (2012); Alexander and Yacoumis (2018).

[10] Sorman and Giampietro (2013).

[11] Boulding (1966).

likely that the fabulous increase in our wealth over the last century will have been a very short-lived episode, a temporal flash in the light of our history, made possible by the unrestrained exploitation of fossil energy stocks that we will have squandered in less than two centuries. To fully understand the problem, we need to think about this in the long term, something that human brains are reluctant to do. If we admit that the world population will reach 9.5 billion by 2050 and if we set ourselves the goal of a standard of living that is that of an average American today—it would be philosophically and politically difficult to legitimise the maintenance of national inequalities—, it is 16 times the current energy that we would need to have: a goal that is not only unattainable but dangerous.

It is time for humans to "come back to Earth" and reasonably and lucidly accept the fundamental unsustainability of the trajectory they are following. Regardless of the inaccuracies of the projections from current models, regardless of the divergences regarding the technical possibilities of producing cleaner and renewable energy, it is obvious that one would have to be very credulous to believe in the compatibility of infinite growth in a finite world. We need to think in terms of magnitude to grasp the utopia in which we indulge. As economist T. Jackson[12] showed in 2009, if we wanted to maintain the same level of growth as that which prevailed over the last 50 years, our GDP would have to be 80 times larger in 2100 than it is today! The technical improvements we can hope for in the exploitation of renewable energies and the efforts made to preserve our environment are simply derisory compared to the implications that constant growth would have on our environment and our energy consumption. Many admit this obvious fact but believe that by then, we will have time to correct our trajectory. This is a gross mistake because the ecological and energy limits are already being reached or are about to be. It is also the classic response of anyone dependent on an addiction that will need to be broken... later.

In fact, our current economic system is doomed to disappear and its disappearance can occur in two different ways. Either we continue our quest for infinite growth, we maintain it by force for as long as our ingenuity will allow, with the inevitable consequence of the destruction of our ecosystem. In this case, the limitation will be ecological because our economies will not be able to survive in a too degraded ecosystem. Or, we manage to maintain this growth without it leading to major ecosystemic disorders, but we will then quickly hit the obstacle of energy shortage, which shortage will signal economic decline in another way. The sooner we are able to take into account the real, the sooner we will be able to face it and the less the social and political consequences will be catastrophic. Any addiction, the addiction to growth and consumption is no exception, is all the less harmful as it is recognised and measures are adopted early to free oneself from it.

[12] Jackson (2009).

4.3 The Cornucopian Fable: Green Growth

Scientific progress has greatly improved our living conditions and those who question it without nuance would be well advised to closely study the living conditions that were ours just a few centuries ago. The rejection of the Enlightenment, the rejection of scientific rationality, the rejection of the very concept of progress, the nostalgia for a lost paradise and a forever magnified past will be of no help in solving current environmental problems. Things are a bit more complex and subtle than that. Science and technological progress are probably necessary to solve the environmental crisis, but they are absolutely not sufficient. They are even potentially dangerous in the current ideological and cultural context.

The cornucopians believe that technological and scientific progress will allow humanity to continue to grow on the small volume that constitutes the earth. In a certain way, they are anti-Malthusians since they consider that no material limit can put an end to growth. They are therefore radical optimists whose extreme version is that of transhumanism. In its most moderate and reasonable form, the cornucopians are strong supporters of green growth or sustainable development. They make decoupling the alpha and omega of what must be implemented to ensure constant growth without endangering the ecological balance of the planet. By decoupling, we mean the possibility offered by technology to produce and consume, therefore to grow, without this having an environmental impact or an acceptable environmental impact. For example, technological innovation in the automotive industry has allowed a car to travel 100 km with 4 litres of petrol when a car of equivalent volume and mass consumed 12 in the past. If we generalise decoupling to all our productive and consumer activities then the ecological impact should grow less than GDP. Overall, it is therefore a question of escaping the observed correlation between GDP, growth and environmental impact. We owe to the biologist Paul Ehrlich[13] an undisputed law useful for identifying the issue and the difficulty of such decoupling. According to this law I = P. A. T where I corresponds to the environmental impact, P to the size of the population, A to the level of wealth expressed in income per individual (affluence in English) and T to the technological factor which conditions, according to its level of efficiency, the impact associated with each unit of wealth. In clear terms, the ecological impact (I) is physically and mathematically linked to the number of individuals on the planet (P) and to their level of wealth and therefore consumption (A). But this impact is modulated by the level of technological efficiency: if the technology is very efficient, the impact will be lower for the same values of P and A than if the technology is inefficient. Decoupling is successful when the theoretical increase in environmental impact caused by demographic growth and wealth increase is perfectly neutralised by the gain in technological efficiency. Everything is fine as long as the T factor compensates for the increase in P and A, necessary for the proper functioning of the economy. Given the constant increase, and desired by economists, of P and A, decoupling must be increasingly

---

[13] Ehrlich (2000).

efficient. Is this possible and has it worked throughout these last 50 years of strong growth and fabulous technological innovations? The answer is clearly negative.

The first reason still lies in the endless race that is being played out here. A race lost in advance, requiring that technological innovation manage *ad infinitum* to guarantee a form of stability and ecological balance where demographic growth and consumption growth generate the imbalance necessary for our economy and our society: growth. Theoretically, it is difficult to imagine what technological miracles would allow to counterbalance not a punctual increase in growth but an infinite increase in growth. Technological innovation cannot constitute a parade because it is physically constrained while our insatiability in terms of wealth and growth is a bottomless pit. The second and main reason for the failure of decoupling lies in what is called the rebound effect or the Jevons paradox named after the economist who identified it. It is a paradox in the sense that the energy gain and the reduction of the ecological impact obtained thanks to a new technology frequently produces an increase in energy expenditure.[14] Jevons, at the end of the nineteenth century, had observed that the invention of the steam engine by Watt, much more energy efficient than the previous device, had actually led to an increase in coal consumption. Precisely, in 1920, the energy efficiency of steam engines was 36 times higher than that of the late eighteenth century. Century, but as the energy produced by steam has increased 2000 times in the same period, the rebound effect has been considerable. For most technological innovations, the rebound effect is around 50%. Since then, the phenomenon has continued to repeat itself to the point of constituting an almost-law. Here are some notable examples.

In the field of communications and digital technology, processors double their efficiency approximately every 18 months.[15] Such an increase in efficiency is theoretically capable of reducing the energy consumed for a fixed amount of digital processing. Yet, energy consumption related to this field is increasing by 12% per year.[16] In fact, this efficiency has allowed for an increase in digital usage, reducing its production cost, leading more people to use it, the price of the megabyte having dropped nearly a million times in 30 years. Another quite singular example is that of lighting. In comparison to candles, oil lamps, gas lamps and other old lighting devices, current bulbs are 700 times more efficient,[17] allowing for a considerable decrease in energy consumption. However, this is not the case in reality. How many individuals now leave a house entrance or leisure areas lit due to the reduced consumption of LED bulbs? In reality, an average French person uses 6500 times more Lumens than their ancestors did when lighting homes became common practice. The technological gain has boosted our energy consumption considerably and this can be systematically observed in the most ordinary daily behaviour: the energy improvement of heating leads to raising the temperature of the house, the reduction

[14] Sorrell et al. (2009).

[15] Koomey et al. (2011).

[16] Vereecken et al. (2010).

[17] Tsao and Waide (2010).

in car consumption leads to more travel, the technological innovation of the TGV has increased the distances between workplace and place of residence, using solar energy for hot water has increased our hot water consumption…

In reality, the rebound effect is a completely predictable phenomenon given the characteristics of human psychology. For the reasons indicated in the previous chapters, our drive to consume will always take advantage of the opportunity given to consume a little more. This can be the cause of a direct rebound effect as in the examples given above. But the rebound effect can also be indirect. This is the case when the savings made in one area, for example the savings related to a more efficient heating system, is used in another equally polluting consumption sector such as tourism travel. The rebound effect can also result from a form of moral alibi such as people who have occasionally adopted a "green" behaviour—for example by acquiring an electric car or by recycling their waste—will develop very unecological behaviours that they would not have had without this form of moral legitimisation.

What about the global level, especially in terms of CO_2 emissions, the main cause of global warming? The analysis of global GDP, energy consumption and CO_2 emissions over the last 30 years is very clear. GDP has continued to increase, as has energy consumption and CO_2 emissions. Admittedly, CO_2 emissions have not increased in proportion to the increase in GDP due to a slight decoupling estimated at 0.7% per year, but they have continued to grow at a level not compatible with the most modest objectives set at the various IPCC (International Group of Experts on Climate) conferences. Clearly, decoupling has not even managed to offset the sole effect of population growth. In fact, CO_2 emissions increased by 40% between 1990 and 2009.[18] More prospectively, according to the IPCC, the level of CO_2 should not exceed 450 ppm in 2050 if we want to contain within bearable limits the extent of global warming. This would require a decoupling of the order of 7% per year with a very low growth scenario of the order of 1.4%, or 10 times more than what we have managed to do so far.[19] With a more ambitious scenario and the goal of generalising growth to all countries, it would require the level of decoupling to be 130 times higher than it is today, a value absolutely out of reach. In reality, containing global warming while maintaining the current growth rate would require no longer emitting any carbon by 2050.[20] This is quite staggering, probably the reason why there are now countless projects as ambitious as they are far-fetched in eco-engineering (seeding the oceans in iron, carbon sequestration, Earth cooling through physical modification of space characteristics, sulphur injection into the atmosphere, creation of artificial ice...) which reveal at once our fetishistic attachment to growth, our panic in the face of the natural limits we are bumping into, our inability to think differently about our society and our vain hope of a technical resolution to a problem that is primarily of a political and psychological nature.

[18] Jackson (2011).

[19] Jackson (2011).

[20] Durand (2018).

In 1971, the economist and Nobel laureate Simon Kuznets developed a very optimistic conception of the capitalist model. According to this author, not only was the level of inequality supposed to decrease once growth had become entrenched and environmental problems should also be resolved because the richer individuals become, the more they would become demanding in terms of the environment: the poor would be in a way "too poor to be green"!!! 50 years later, it is clear that none of these predictions have come true. Kuznets and all the cornucopians are, it seems, victims of magical thinking or, to put it another way, they take their desire for reality. The orders of magnitude presented in this chapter, whether it is about growth, environmental impact, CO_2 production, decoupling capacity, do not allow us to imagine that the current economic model can survive the ecological crisis regardless of the level of technological innovation considered. Such a system is running out of steam. It naturally produces major social inequalities, between individuals and between states, likely to generate major social unrest and, in an attempt to contain them, the system must raise the global wealth level through infinite growth. This infinite growth is of course a myth, partly because of the physical and energy limitations of our environment, and partly because of the global environmental destruction that results from our growth and which, in the long run, could lead to some form of collapse. Faced with this aporia, the system is completely helpless since it is perfectly impossible to simultaneously satisfy two incompatible objectives which are the growth necessary for social peace and the preservation of the planet: green growth or sustainable growth is just a pleasant and friendly myth but not realistic and, in the end, mystifying. A technically unrealistic myth because the race for infinite growth has no technical solution in a finite world and psychologically unrealistic, because as soon as a technological efficiency gain is achieved, it will be immediately offset by the power of the rebound effect. The slogan and cliché "not consume less but consume better" illustrates better than all the others, our ability to delude ourselves with a rhetoric as hollow as it is unfounded: astonishing capacity of the human brain to develop clever sophisms to preserve the status quo. If the cornucopian fable still wins the adherence of some, it is of course because it allows us to believe that the system can survive and that surface adjustments, mainly technological, will ensure its continuity: nothing is more difficult for humanity than to change the programme that has led it to where it is. This is where lies the colossal strength of the economic and political system in which we live. Growth is necessary at the macroeconomic level, but it is also, naturally, the objective that each individual pursues: very few must be the individuals whose stated goal is to impoverish themselves. It is the conjunction and convergence of forces that operate at the individual and collective level that make the system almost unalterable, at least for the moment and probably still for a few decades. The degrowth movement is undoubtedly the most rational response given the speed at which we are approaching the precipice. Although this movement has gained a tiny form of recognition recently after being heavily criticised by liberal economists, it remains in the minds of many a form of revolutionary utopia totally disconnected from reality. In a certain sense, this is not false. Degrowth is a rational solution, perhaps too rational, and probably unrealistic because of the intrinsic irrationality of Man. Yes, degrowth, or stable growth, is the solution: it is even the only solution. But it is probably not yet operational because the ordinary and short-term interests of individuals oppose the implementation of a relationship with the world that breaks with a few thousand years of tradition and which is hardly compatible with the psychological forces that animate us all.

Faced with such a situation, it is tempting to look for and identify the scapegoats, in other words those who are at the helm of our society, those who have the power to decide, those who legislate, those who, through their financial power, have the ability to steer the course of the world. Of course, it is rare that when one wins at a game, one agrees to change the rules. What political leader, what great business leader, what human at the height of personal success would agree to change rules that are very favourable to him? Probably none, as it is likely that they are all deeply attached and dependent on the privileges they have acquired or that have been granted to them because of their birth. However, are we sure that the others, the ordinary citizens, would be ready to change the rules of the game? I believe I have partly shown that they would not. Who would consent to less comfort and less wealth to avoid the punishment of the precipice? This is the whole complexity of the situation. How many of us vilify Google while using it several hours a day, to take just one example among thousands of others? We are relatively powerless elements of a complex system that is beyond us, that we contribute more or less voluntarily to keep alive, that we often perceive as harmful, but on which we are terribly dependent. We are dependent on it as it is dependent on us. To fully understand this alienation, we now need to understand its historical origin.

References

Alexander, A., & Yacoumis, P. (2018). Degrowth, energy descent, and 'low-tech' living: Potential pathways for increased resilience in times of crisis. *Journal of Cleaner Production, 197*, 1840–1848.

Boulding, K. E. (1966). The economics of the coming spaceship earth. In H. Jarrett (Ed.), *Environmental quality in a growing economy*. Resources for the Future/Johns Hopkins University Press.

Büchs, M., & Koch, M. (2019). Challenges for the degrowth transition: The debate about wellbeing. *Futures, 105*, 155–165.

Costanza, R., Daly, L., Fioramonti, L., Giovannini, E., Kubiszewki, I., Mortensen, L. F., Pickett, K. E., Ragnarsdottir, K. V., De Vogli, R., & Wilkinson, R. (2016). Modelling and measuring sustainable wellbeing in connection with the UN Sustainable Development Goals. *Ecological Economics, 130*, 350–355.

Durand, F. (2018). Global warming: Colossal stakes for new societal projects. *Parcours*. Cahiers du GREP Midi-Pyrénées, GREP MP. hal-02004842.

Ehrlich, P. R. (2000). *Human natures: Genes, cultures, and the human prospect*. Island Press.

Gowdy, J. (2020). Our hunter-gatherer future: Climate change, agriculture and uncivilisation. *Futures, 115*, 1–9.

Jackson, T. (2009). *Prosperity without growth: Economics for a finite planet*. Earthscan.

Jackson, T. (2011). Societal transformations for a sustainable economy. *Natural Resources Forum, 35*, 155–164.

Kallis, G., et al. (2018). On degrowth. *Annual Review of Environment and Resources, 43*, 4.1–4.26.
Koomey, J. G., Berard, S., Sanchez, M., & Wong, H. (2011). Implications of historical trends in the electrical efficiency of computing. *Annals of the History of Computing, IEEE, 33*(3), 46–54.
Moriarty, P., & Honnery, D. (2012). What is the global potential for renewable energy? *Renewable and Sustainable Energy Reviews, 16*(1), 244–252.
Sorman, A. H., & Giampietro, M. (2013). The energetic metabolism of societies and the degrowth paradigm: Analysing biophysical constraints and realities. *Journal of Cleaner Production, 38*, 80–93.
Sorrell, S., Dimitropoulos, J., & Sommerville, M. (2009). Empirical estimates of the direct rebound effect: A review. *Energy Policy, 37*, 1356–1371.
Tsao, J. Y., & Waide, P. (2010). The world's appetite for light: Empirical data and trends spanning three centuries and six continents. *Leukos, 6*, 259–281.
Vereecken, W., Van Heddeghem, W., Colle, D., Pickavet, M., & Demeester, P. (2010). Overall ICT footprint and green communication technologies. Paper presented at the 4th international symposium on communications, control and signal processing, Limassol, Cyprus.

Chapter 5
The Strength of the Group and the Weakness of the Individual: Views from Anthropology and Evolutionary Psychology

Abstract In this chapter, I draw on work conducted in anthropology and evolutionary psychology and on the importance of the Neolithic revolution. During this revolution, humans formed increasingly large groups and gradually transitioned from a subsistence economy to a production economy. The groups thus formed, which could number hundreds of thousands or millions of individuals, imposed a new way of relating to the world from which no human can truly escape. The individual owes their survival to the group but must also submit to the group's law. They thus lose some of their autonomy and are forced to compete with other members of their community for survival, thereby contributing to economic growth, which we have seen is antithetical to the preservation of the planet.

In the preceding chapters, we have presented a multi-level causal chain, partly explaining the current environmental crisis. We now need to go further in justifying this analysis. The main aim will be to answer two questions. The first is to explain how humans have managed to dominate the planet so remarkably as to alter its biophysical balance and endanger their own survival. The second will be to ask, for what other reasons than those presented in the previous chapters, we are unable to react rationally and break with an economic mode of operation that is both unsustainable and deeply unequal.

Undoubtedly, the reader more trained in the humanities than in the biological sciences will be surprised that I invoke biological and psychological determinisms as explanatory factors of human behaviour at the political and economic level. I would like to make two preliminary remarks on this subject. Obviously, I do not deny the essential part that culture plays in the evolution of our societies. But our cultures are also dependent on our biological and evolutionary history in a very Darwinian sense of the term. Fortunately, cultures are also capable of influencing the raw forces of biological nature that constrain us. What I would like to show here

T. Ripoll, *Human Psychology and the Destruction of the Planet*, Anthropocene – Humanities and Social Sciences, https://doi.org/10.1007/978-3-032-02080-2_5

is how the biological forces resulting from the evolutionary processes that led to the advent of our species are precisely those that are leading us today to the brink of the precipice. Let us never forget that our modern *Sapiens* brain is almost no different from the hunter-gatherer brain we still had 15,000 years ago. It would therefore not be surprising if what was adaptive in the cultural and biophysical environment of the Palaeolithic is no longer so in the current environment. Understanding our distant history, understanding the biological forces that shaped it, and understanding its cultural evolution is the best way to identify the reasons for our inability to manage the crucial problem that is posed to us today.

5.1 Not So Long Ago… Human Was an Animal Almost Like Any Other

Not so long ago, just over 10,000 years ago, we were still hunter-gatherers. So we have been hunter-gatherers for 99.5% of the history of the genus *Homo* and for 95% of the history of the species *Homo Sapiens.* There is little doubt that this distant past still plays an important role in our current behaviour. During the 300,000 years of our *Sapiens,* history, evolution has been almost exclusively cultural and almost no longer biological. We are genetically Stone Age men because biological evolution operates only over long periods of time, cultural evolution over short periods. But even cultural evolution was incredibly slow for most of the 300,000 years of our species' existence. From generation to generation, children, grandparents, great-grandparents knew exactly the same environment, had the same modest technology and shared a culture that did not vary from one generation to the next. Nothing like the difficulties that parents today face in understanding even the mental universe of their own children, and slightly older adults in assimilating the technological and societal revolutions that punctuate at an ever-increasing pace the temporality of a simple and so brief human life. The landscapes that were theirs did not evolve. Only the regular migrations they undertook to ensure their survival or already to quench their thirst for discovery led them to see other horizons. At that time, we were a species like any other, in small numbers: a rare, very rare species—about 15,000 individuals on French territory in −20,000[1]—compared to the abundant fauna of the time and which could easily have gone unnoticed. A species that certainly experienced major demographic regressions caused by the numerous phases of climatic upheaval of the Pleistocene. The human population was very stable for nearly 300,000 years, not exceeding six million until 10,000 BC.[2] Its growth rate was almost zero and often negative. We then lived in small tribes probably not exceeding a few hundreds of individuals who mostly had kinship ties. Archaeological studies and anthropological studies conducted on the last hunter-gatherers of the early twentieth century have identified characteristics of these societies that break very much with the Rousseauist conception of the "noble savage".

[1] Dortier (2015).

[2] Biraben (2003); Bocquet-Appel (2011).

First, our distant ancestors had no kind of ecological consciousness and had no other form of respect for nature than that of preserving it when necessary for their food and survival.[3] Everyone knows the famous phrase of the Amerindian chief Seattle, "Every part of this Earth is sacred to my people… the Earth does not belong to Men, Men belong to the Earth", often invoked by a nostalgic ecological current of a time of perfect and mythical harmony between human and nature. In reality, although human was equipped with rudimentary weapons and although his population was very low, he was already a formidable predator. The most contemporary research leaves no doubt: numerous species of the Pleistocene's large fauna disappeared due to anthropic pressure and not, as previously thought, solely due to climatic upheavals.[4] These extinctions generally occurred when humans colonised new territories, particularly biogeographically isolated territories such as islands or continents devoid of any human presence. In Africa, where hominids have been present for several million years, the wildlife has had time to adapt its behaviour to the danger of humans, whereas in South America or Australasia, lands of more recent conquest, the fauna has not been able to evolve to escape this super-predator. This is why in Africa only 18 species of mammals have disappeared during our prehistory, while this number rises to 25 in Australasia and 62 in South America. The large wildlife of the vast expanses of North America is in fact only a pale reflection of what existed when the Native Americans explored and began to exploit these territories. Humans did not wait for firearm technology to make large fauna disappear.

Our distant ancestors were not the gentle pacifists we often imagine. Violence was frequent and at a level that is difficult for us to conceive of today. As with current chimpanzees, this violence was sometimes exercised between close tribes due to competition for resources necessary for their sustenance. It was mainly exercised within a group. The main cause of intra-group violence remained the struggle for status and, but this is generally related, the competition to obtain the sexual favours of women.[5] Competition for food resources and competition for the possibility of reproduction constituted and still constitute the two main motives for lethal conflict. As a Yanomami Indian noted, perhaps humorously: "even if we love meat, we love women even more".[6] Sexual rivalry was all the more violent as polygyny was universal and female infanticide was the rule,[7] leading to a significant demographic imbalance. Among the Eskimos, the homicide rate was estimated at 1 in 10,000 per year, ten times more than the American homicide rate, which is nevertheless the highest among Western countries. Among the Australian Aborigines, 30% of the adults of the Murngin tribe died a violent death. Among the Yanomamis of Brazil, 15% of adults died due to internal and external group conflicts. The record is held by the Waorani Indians of the Amazonian Ecuador: 60% of adults disappeared due to violent conflicts. Only the most violent, strongest, and undoubtedly the most

[3] Smith and Wishnie (2000).

[4] Sandom et al. (2014).

[5] Gat (2006).

[6] Chagnon (1977).

[7] Divale and Harris (1976).

intelligent and socially adept men could survive. We are their descendants and this fundamental aggressiveness is still painfully restrained by the frameworks and institutions we have laboriously put in place. Of course, we should not therefore succumb to the cliché of prehistoric human as a naturally violent brute. The reality probably lies halfway between Rousseauist angelism and the miserabilist conception of nineteenth-century evolutionary anthropology, according to which human is naturally violent.[8] During the Paleolithic, intraspecific aggressions were probably frequent, but violence was not organised:[9] it was still violence without war.

As with most mammals, this mortality mainly affected Men—women were frequently kidnapped and raped during victorious raids—and was particularly pronounced among young adults at the peak of their reproductive and physical capacity. Risk-taking and violent acts are still very much present among young men today: the peak of male violence is reached around 25 years of age in most of our modern societies.[10] It is also at this age that men are most ready to leave their family and their country when local conditions do not allow them to obtain the social status they covet, nor to find a sexual partner likely to allow them to start a family. Competition for sex and competition for power were intimately linked, as evidenced by the strong correlation between the two, as is observed in most higher animals. We have little reliable data on this subject, but it is all convergent. The data from the first ethnological research conducted on the Aborigines of Australia, before their way of life was affected by the European conquest, is a remarkable illustration of this. In the arid centre, a powerful man could have 5 or 6 wives when the less socially powerful had none. In the richer and less inhospitable regions of the North, men of high social status had a dozen wives. Exceptionally, the great chiefs had more than double that. This example, confirmed by many others, reveals the existence of a strong relationship between "sexual wealth" and the density of environmental resources. When a society is able to produce more wealth due to a more favourable environment, inequality in the number of sexual partners increases. In hostile environments that hardly allow for the accumulation of wealth (the desert, arctic zones…), social differentiation is less marked and inequality in terms of sexual partners is almost nil, which is why monogamous marriage is the norm among the Eskimos.

Another important characteristic of Paleolithic tribes and those that survived into the twentieth century is their "violent" egalitarianism, in the sense that any attempt to appropriate wealth was prohibited or punished.[11] Studies conducted among contemporary hunter-gatherers have revealed that the group could kill the one who, through selfish or aggressive behaviour, tried to appropriate more resources than the others.[12] The distribution of wealth, of prey or harvested plants, followed fairly strict

[8] Majolo (2019); Wrangham (2018).

[9] Pathou-Mathis (1993).

[10] Jones (1993).

[11] Boehm (2013)

[12] Boehm (1999).

sharing rules.[13] Such an ethical imperative was simply adaptive. In a society that did not accumulate and did not produce more than it could consume, an unfair share would have led to the death of those who would have been left behind. The egalitarianism of pre-Neolithic societies is an uncontested fact[14] although it may seem contradictory with the violent competition I described earlier and with an inequality in terms of sexuality which we have seen is correlated to the social status of individuals. The level of egalitarianism could vary from one culture to another and it was all the more marked as environmental conditions were difficult and did not allow for the accumulation of wealth. Nomadism also contributed to this as it is difficult to carry precious goods on the backs of men. In reality, social differentiation, that which is likely to give power and contribute to sexual wealth, was more symbolic than material, which is why powerful chiefs have probably for a very long time adopted some singular mark (coloured feather, various and varied ornamentation, tattoo…) signalling and consecrating their status as we do today through ostentatious consumption. But within a small Paleolithic community, as within the last hunter-gatherer communities, wealth inequalities remained tiny or non-existent, just as we observe in higher social animals. A dominant chimpanzee or an alpha wolf does not have more goods than the rest of the pack. Nevertheless, he remains the holder of a power that results from certain physical or psychological abilities recognised by the members of the pack. It is the power associated with his hierarchical position—social status—, power often acquired at the price of a merciless struggle, which gives him greater reproductive opportunities. This equality of wealth in the Paleolithic was associated with a lack of differentiation in terms of division of labour, except for differences related to age and sex.[15] Each individual of equivalent age and sex was able to perform the functions of any other individual. Another striking point in the society of our distant ancestors is also its extreme stability. Today we cannot even conceive that the society of tomorrow could be identical to that of today. We have so strongly integrated the idea of constant evolution or progress that foresight has become a discipline in itself with its literary avatar: science fiction.

In a certain way, the lifestyle of our ancestors, at least in terms of their interactions with their ecosystem, did not differ much from that of other higher mammals. Certainly, their intelligence and their capacities for coordinated action led them to exterminate many species, but their overall impact on the Earth's ecosystem remained almost nil. Everything will suddenly change with the appearance of agriculture and livestock.

[13] Turchin et al. (2013).

[14] Knauft (1994).

[15] Gowdy (1998).

5.2 When Human Distinguished Himself from Other Animals: The Neolithic Revolution

The human trajectory throughout its history has never been linear. It has been marked by ruptures and cultural revolutions which have, sometimes over relatively short periods, totally renewed our existence as well as our relationship to the world. For many anthropologists, biologists, economists, psychologists, the Neolithic revolution was the most important of them: the one that made human an animal not quite like the others and which foreshadowed contemporary society. From my point of view, this revolution has inexorably led us to the Anthropocene and therefore to the current ecological crisis.[16]

The emergence of agriculture and livestock farming has completely changed the relationships between humans and the relationships between humans and their environment. The incredible increase in population is probably one of the most visible consequences of this major revolution. While the demography has been stable for nearly 300,000 years, the human population has gone in a few thousand years from six million—it was about 1 to two million 50,000 years ago—to 200 million[17] once agriculture and livestock farming became widespread. What is extraordinary about this phenomenon is that it occurred at roughly the same time in 7 or 8 geographically very distant areas without any possible cultural exchanges between these societies.[18] When Cortez landed in Mexico in the sixteenth century, he discovered other humans whose existence was not even suspected in Europe. But he discovered above all roads, palaces, schools, irrigation canals, markets, craftsmen, temples, arts, books, music, judges, priests, an administration… In short, in a slightly different form, he found his own culture some 10,000 kilometres from his own country. The impression must not have been very different from what we would feel if we discovered individuals with a culture similar to ours on another planet. The simultaneity of the Neolithic revolution remains a major subject of scientific research, but what is certain is that it cannot have resulted from chance. It is largely explained by the establishment of favourable and stable climatic conditions of the Holocene. The abrupt climatic variations of the Pleistocene could not allow the establishment of a developed and systematised agriculture.[19] It is also probably explained by the existence of evolutionary forces that led to the emergence of universal psychological properties—cognitive and social—that could only lead to the Neolithic revolution: on the one hand exceptional cognitive abilities, on the other hand extreme sociability.

Contrary to what common sense considers, the invention of agriculture did not improve our living conditions.[20] Yet, it was the starting point of a dynamic that led

[16] See Gowdy and Krall (2014) for a detailed analysis of this evolution that led us to the Anthropocene.

[17] Biraben (2003).

[18] Wright (2004).

[19] Gowdy and Krall (2014).

[20] Larsen (2006).

to the unchallenged domination of human over the planet and to the now completed disappearance of the hunter-gatherer lifestyle. This is a paradox that needs to be explained and that could be summarised as: "misery for the individual and power of the group". For although agriculture and livestock farming initially worsened the living conditions of humans taken individually, they gave them unprecedented power at the level of the group they constituted. In fact, life expectancy did not increase during the Neolithic revolution. Overall, the health of Humans even declined significantly.[21] To take only the example of France, life expectancy in the middle of the eighteenth century was still 25 years. It is only very recently, in the last 150 years, that life expectancy has returned to a level that was that of the Pleistocene hunter-gatherers, estimated at 33 years.[22] Something similar is found at the level of anthropometry. It seems that humans lost about ten centimetres during the transition from the Palaeolithic to the Neolithic. It is only since the industrial revolution that the size of humans has returned to the level it reached before the Neolithic. As size is very dependent on living conditions and more specifically on the quality of food, this is another indicator of the overall degradation of human health.[23] In fact, nothing allows us to suppose that the diet improved: it changed. There was simultaneously an increase in the consumption of cereals and a decrease in the consumption of wild animal meat. This has contributed to the appearance of new deficiencies, absent in the Palaeolithic.[24] A very large study carried out on skeletons from before and after the Neolithic revealed pathologies specific to the work of the land, notably bone and joint pathologies such as arthritis resulting from the intensification of a specific work to which Humans were not adapted. On the other hand, with the emergence of agriculture, the first large cities appear. 10,000 years ago, the city of Jericho had 2000–3000 individuals spread over an area of 2–3 hectares, much more than the limited number of individuals in a group nomadic or seminomadic of the Palaeolithic. This "urban" concentration, which was rapidly intensifying, led to the spread of epidemics for which there was no prophylaxis at the time. In terms of what we would call quality of life today, there is also no evidence of any improvement. With the demand for productivity and the time constraint linked to the agricultural production cycle—periods of ploughing, sowing, harvesting and processing—humans were forced into phases of intense activity, probably all the more exhausting as the technology remained very archaic. Conversely, hunter-gatherers, strategically moving in game-rich areas, spent relatively little time searching for their food and had much more leisure and socialisation time than ours today.[25] Let's remember that during the industrial revolution, adults often worked more than 80 h a week when the time imposed by food search is estimated at less than 5 h a day for most hunter-gatherers. Moreover, the compara-

---

[21] Larsen (2006); Cohen and Crane-Kramer (2007).

[22] Kaplan et al. (2000).

[23] Hilt and Formicola (2008).

[24] John Gowdy and Baveye (2019).

[25] Sahlins (1968).

tive productivity of hunter-gatherers and early farmers exploiting their own physical strength was clearly in favour of the former.[26] The inequality introduced by the Neolithic revolution also largely contributed to the deterioration of the quality of life of a very large number of individuals. Until the nineteenth century, living conditions were deplorable in rural Europe. The diet was very poor and famines were frequent. Simply having a roof and being able to dress were goals that not everyone achieved. In the cities, the situation was even worse.[27] In Florence, during the Renaissance, statistics reveal that 80% of the population was considered very poor or indigent. In deprived neighbourhoods, 10 people lived in the same room and a dozen families were crammed into unsanitary houses without even running water. Under Louis XIV, 10% of city dwellers were beggars and 30% were so poor that they could not feed themselves normally. Conditions were so harsh in Europe that many parents abandoned their children who were taken in by orphanages. In Paris, at the end of the eighteenth century, 8000 children were abandoned each year. Most remained illiterate and uneducated, vulnerable and therefore shamelessly exploited by the emerging industry. A real trafficking system was set up to provide almost free labour to the clothing factories in Belgium and France. In short, in a few thousand years, the Neolithic revolution had not at all improved living conditions, except for the tiny minority of those who were able to enrich themselves at the expense of the poorest. The miserable fate of the billions of humans surviving in the large Asian or African megacities today is only the catastrophic extension of the misery also engendered by the Neolithic revolution.

So why, if agriculture and livestock farming did not make life easier for humans as a whole, did the Neolithic revolution spread across the entire planet? First of all, it should be remembered that long before the Neolithic revolution, humans mastered agriculture. Traces of sorghum cultivation have been found in caves occupied by humans 100,000 years ago in Mozambique. Even if this discovery is disputed, there is no doubt that humans had perfectly grasped the possibility of sowing and producing consumable plants, which they did in a rather unsystematic and marginal way long before the Neolithic.[28] They had also already domesticated the wolf to make it a companion and an ally in hunting: the dog. Nothing stood in the way of the domestication of sheep and cattle, which only appeared 20,000 years later. The transition from the Palaeolithic to the Neolithic therefore took place gradually. There was not a sudden discovery that would have changed the lives of humans. Rather, the lives of humans would have changed gradually to the point where at a certain moment, the lifestyle of hunter-gatherers was no longer possible even though it provided many advantages at the individual level. As is often the case, it is the conjunction of environmental factors—climate, distribution of exploitable wild grasses, seasonality, moderately abundant wild natural resources—and cultural factors that explains this type of rupture. It is not possible to synthesise it here. But

---

[26] Bowles (2011).

[27] Ponting (2007).

[28] Zvelebil and Rowley-Conwy (1986)

what is certain is that by developing agriculture and livestock farming, probably initially limited, humans gradually moved, over a few thousand years, from a subsistence economy to a production economy: a dynamic that, once started, could only strengthen. The fundamental turning point occurred when this production no longer had the sole purpose of being all consumed immediately. In doing so, it could be used to cope with the less productive periods of winter and spring in temperate regions. Human had stopped hunting or gathering for his present needs; he was producing and preserving for his future needs. The cognitive and cultural ability to think about production with the aim of future use has was probably a major determinant of this revolution. Since there was production beyond immediate needs, there had to be storage, the latter implying its control, protection and management. Storage, necessarily close to the cultivation areas, also implied sedentarisation. Because of this sedentarisation, it is likely that hunting resources decreased because hunting or fishing in the same territory for too long inevitably leads to a scarcity of prey to the point where they could no longer meet basic food needs. As a result, humans became increasingly dependent on agriculture, which they had to systematise, perfect and intensify, with hunting and fishing becoming marginal activities. Sedentarisation further increased the need to accumulate to cope with possible shortages linked to climatic hazards or the seasonality of production. It finally allowed accumulation with the prospect of personal enrichment, which was hardly possible in the case of nomadism. Whereas with nomadism, social status could not depend on accumulated material wealth, the first breeders and farmers probably quickly identified the social advantage that could represent the accumulation of productive surfaces or animals. Production could gradually become an end in itself and all human energy and ingenuity had to be deployed to increase the level of production. The search for continuous growth thus began to emerge. The goal of improving productivity must have gradually colonised minds, definitively breaking with the cultural and technical stability of the Palaeolithic. The concept of progress probably emerged at this distant time. It had no reason to exist before. At both the individual and collective level, the increase in production became both a guarantee of individual and collective survival. It also gave, at the individual level, a possibility of personal enrichment that was not even imaginable for hunter-gatherers. The demographic impact was just as important. Groups of hunter-gatherers of thousands or tens of thousands of individuals would have had no chance of survival. All animal species, especially social predatory species, are very constrained demographically. Orca groups rarely comprise more than 15 individuals, just as wolf packs have fewer than a dozen individuals in Europe. Below or beyond, the group would be less efficient in its predatory activity. Human is subject to the same ecological constraints, which is why they rarely formed groups of more than 150 individuals before the Neolithic. Agriculture and livestock farming radically changed the game. No ecological barrier, except for food shortage, could limit the demographic increase of these communities. The increase in the number of individuals further distanced these first rural populations from the possibility of maintaining a hunter-gatherer lifestyle. Without really choosing it, it is therefore a whole chain of punctual, cultural and environmental processes that have gradually transformed the human and made him the farmer-breeder still so present in our culture. The logic of small steps has definitively changed the face of humanity.

The question that arises is why almost all of humanity was won over by the Neolithic revolution when at the individual level, this new way of life neither facilitated nor made life more pleasant. To understand this, it must be remembered that humans are subject to individual competition within their group of belonging and to collective competition between different groups. It is at this level of competition that the domination of farmer-breeder groups over hunter-gatherer groups was played out. The Neolithic transition has also been called the ultra-social transition in the sense that if humans have dominated the world from this time, it is certainly because of their remarkable cognitive abilities but especially because of their extreme level of sociability, greatly accentuated by the new way of life they had just adopted. We are powerful only thanks to the combined capacities of a large group of humans. It would be enough to isolate a small group of humans for a few generations for this group to regress culturally and barely distinguish itself from other animals in its environment. What would we become and what would we be able to produce and do if we were isolated with a hundred of our relatives on an island isolated from the rest of the world? Such an experiment occurred when small groups of Australian aborigines, after reaching Tasmania, were, until Western colonisation, isolated from the rest of the world due to the rise in oceans that occurred 10,000 years ago. When the Europeans met them, they had almost lost all of the rich Australian aboriginal culture from which they originated and used only 24 tools, all extremely rudimentary. Their number was too small to maintain the initial culture.[29] They were very quickly exterminated. The cognitive anthropologist, Joseph Heinrich, has perfectly described and even mathematically modelled the social and cognitive processes that allow a culture to develop, maintain or regress.[30] The conclusion is clear. The larger the human community, the richer the culture and the more likely it is to gain in complexity. When the human community is demographically reduced and exchanges between communities are rare, the culture cannot evolve and tends to regress in case of isolation of too few individuals.[31] It is indeed from their ultra-sociality that humans derive their superiority. It is also this phenomenon that led to the domination of Neolithic culture over Paleolithic culture. The mere fact of establishing large communities of men, from villages to cities, led to a large-scale cultural sophistication giving the farmer-breeders a formidable technological and then military superiority. As, moreover, this new way of life is by nature expansionist and conquering, the disappearance of Paleolithic culture was absolutely inevitable. The competition between human groups of different culture was fatal to hunter-gatherers whatever the disadvantages, at the individual level, of adopting such a new way of life.

[29] Diamond (1978).

[30] Henrich (2004).

[31] Derex et al. (2013).

5.3 A Big Step Towards the Anthropocene

The changes brought about by the Neolithic revolution are profound. In this new culture, the constraint of equality has gradually disappeared because the possibility of accumulating wealth quickly appeared as the most effective way to win the intra-group competition. Only societies producing, even if temporarily, more than what is necessary for survival can become unequal: without overproduction, little possibility of differentiated appropriation of resources.[32] It was no longer a matter of hunting or gathering what was necessary for the group's survival on a day-to-day basis. It was about producing and accumulating more in order not only to ensure individual survival but also to assert one's social status vis-à-vis potential rivals. Inequality was no longer fought against. It had become the norm by leading to the practice of inheritance, a practice absolutely inconceivable in hunter-gatherer societies. The obsession with maintaining one's level of wealth and social rank through the coming generations led to the origin of infanticide to avoid dispersing the acquired heritage over several children, especially women. The dynamic generated by the desire to increase one's level of wealth naturally led to episodic surpluses of production. It was at this point that trade exchanges appeared, contributing to the increase in inequalities… already. In a society becoming increasingly complex, the first bureaucracies emerged, on the one hand to frame the nature of these exchanges, on the other hand to legislate, particularly on the establishment of private property and on the rules allowing to ensure social peace in a growing context of inequality. It should not be forgotten that during the Paleolithic, all individuals in a community knew each other and conflict resolution often took place within an extended family framework. For the first time in the history of living beings, an animal had built a society of extreme complexity within which individuals not related by family ties had to cohabit. Strangers had to coexist peacefully and collaborate. Therefore, a whole set of new functions appeared. The division of labour was established, replacing the indistinctness of functions that dominated during the Paleolithic. A real social hierarchy was established with its classic lot of inequalities inherent in any stratified society. It is from this time that slavery appeared.[33] In order to meet the needs of the increasingly numerous individuals, ensuring the new jobs required for the stability of these complex societies, it became necessary to increase agricultural production. The number of "unproductive"—military, priests, bureaucrats of all kinds—continually increasing, it soon became necessary to gain new territories. The first empires were forming, and with them, the first armies financed by surpluses of production that had become indispensable. Inter-group conflicts were becoming more and more frequent requiring ever more resources and soldiers. Violence no longer resulted from individual rivalries. It could oppose large groups of individuals, hence the importance of the status of warrior. It is also at this time that the first large necropolises appeared in which skeletons of men probably died as a result of

[32] Hayden (2008).

[33] Pathou-Mathis (1993).

aggression between distinct groups of humans were found. Only the largest, most structured, best organised, most technologically innovative and most productive cities could survive conflicts that could only occur in an expansionist and conquering dynamic that the Neolithic revolution had imposed. The small nomadic or semi-nomadic groups of the Paleolithic had no chance of survival against the power of Neolithic societies. Broadly speaking, our current society was in its infancy and the essential ingredients of modern capitalism were already present. Only missing were technological development and access to fossil energies that would boost the level of growth and cause the deep destabilisation of our ecosystem that poses a problem for us today.

5.4 From the Autonomous Individual to the Heteronomous Individual

The ultra-sociality of humans was undoubtedly a determining element of its prodigious development. He is the only among all animals to be able to cooperate intensely with individuals he does not know and who are not genetically related to him.[34] According to the theory of the "social brain",[35] the evolution of the brain and human intelligence has been largely guided by the necessity to interact with others, either through competitive relationships or through cooperative relationships. Language would have appeared to make coalitions and alliances, so important in human life, more efficient, in the Palaeolithic as in the Neolithic.[36] Our ability to form very large communities of individuals capable of cooperating, however, has a dark side: organised and large-scale war. Humans, as far back as we can trace their past, have been potentially violent animals. But until the Neolithic, these conflicts were sporadic and at worst involved only a few dozen individuals. It is the extreme sociability of humans and their altruism towards the group they belong to that are at the origin of the emergence of large-scale conflicts. The violence inflicted on individuals outside the group is only equalled by the altruism and generosity granted to individuals of one's own group. This is what is called parochial altruism, another specificity of human psychology consisting in extremely and incoercibly favouring individuals of the same ethnicity, culture, language, religion… Current or past nationalism and racism are clear manifestations of this. The campaign slogan of President Trump, "American first" or that of the French far right "the French first" only exploits the strength of parochial altruism. We believe ourselves to be free individuals when in reality we are socially constrained to a point that is difficult for us to admit. Psychologist Henry Tajfel conducted a very enlightening experiment on

[34] Wilson et al. (2013).

[35] Byrne and Whiten (1988); Dunbar and Shultz (2007).

[36] Tomasello (2009); Richerson and Boyd (2005).

this subject.[37] He asked subjects to express an aesthetic judgement of Klee and Kandinsky paintings. Once this task was completed, he made them believe that he had formed two groups based on their agreement in the aesthetic evaluation task. In reality, he had formed these two groups randomly. He finally asked them to distribute a sum of money between the individuals of their group and the other group. Systematically, the subjects tended to give more money to the individuals of their group. What is remarkable here is that the effect of the group is considerable even though the group is not only purely virtual and abstract, but also formed just a few minutes before the subject operates the sharing of the money entrusted to him. The ultra-sociality of humans makes them excessively dependent on their group of belonging. This powerful social force is of course at work in patriotism and constitutes a major obstacle to the perception of the human species as one and indivisible. Ultra-sociality makes the group powerful, but it harms the autonomy of individuals and their ability to make choices different from those implicitly or explicitly admitted by their group. The Neolithic revolution contributed to accentuating the dependence of individuals on a large group of belonging (the city, the state or the empire) to the point of greatly hindering their decision-making capacity and the possibility of adopting a line of conduct alternative to that imposed by the group.

The very deep analogy with the behaviour of large societies of social insects will allow us to better grasp the importance of ultra-sociality and its exacerbation since the Neolithic revolution. The consequences of the ultra-social transformation of the Neolithic are analogous in social insects and in humans, leading the great biologist Edward Wilson to evoke in these two cases "the social conquest of the Earth".[38] Like us, social insects largely dominate the ecosystems in which they live and are naturally expansionist. In the Amazon rainforest, social insects account for 33% of the total biomass and 75% of the insect biomass.[39] They explore and conquer all ecosystems that are favourable to them. The number of ants in the world is estimated at 10,000 trillion. Although an ant weighs one millionth of the weight of a human, their total biomass is similar to that of humanity: an unprecedented evolutionary success made possible by an ultra-sociality as extreme as that of humans. Just as we are collectively driven by a growth objective, social insects are often driven by a very specific objective. Leaf-cutter ants live in societies of millions of individuals, organised in a complex way with the sole objective of cultivating a fungus necessary for their food. They have, long before us, realised the Neolithic revolution. Leaf-cutter ants are mono-culturalists, but they use all kinds of plants to make an organic manure necessary for the growth of fungi. They take care to eliminate weeds with substances made for this purpose and can exchange the product of their culture with other colonies, sometimes with colonies of different species.[40] The complexity of their society goes hand in hand with a hyper-specialisation of individuals and a rigid

---

[37] Tajfel et al. (1971).

[38] Wilson (2012).

[39] Fittkau and Klinge (1973).

[40] Gowdy and Krall (2014).

hierarchical organisation. Hyper-specialisation leads each individual to perform only one task in a repetitive and very effective way without it being possible to perform other tasks than the one for which he was designed. Unlike humans, however, the roles they perform are dependent on their phenotype (soldiers are physically different from workers or the queen). Their function is genetically defined. For humans, roles are culturally constrained and therefore contestable, although the environment in which they are born massively determines their future hierarchical position. However, a human is not biologically programmed to be a soldier, a farmer, a billionaire or a minimum wage earner. The Indian caste system is nothing more than an unfortunate attempt to apply the hyper-hierarchical social organisation of social insects to humans.

This hyper-specialisation and stratified organisation of society, the exact opposite of Paleolithic societies, is the price to pay for an increase in productivity that allows growth, development, world conquest and the domination or destruction of competing societies. Adam Smith had perfectly foreseen this in "The Wealth of Nations", considering the division of labour as necessary for productivity, growth and economic development. That this specialisation is by nature unequal is a problem for the individual, but not for society, which can continue to grow on the "backs" of its slaves.

Ultrasociality increases collective intelligence, that is, it gives the resulting superorganism an increase in its ability to solve complex problems that an isolated individual could not solve, hence their power and domination over the rest of the ecosystem. But, and this is almost a paradox, this increase in collective intelligence—the intelligence of the superorganism that is our society or any colony of social insects—is achieved at the cost of a decrease in individual intelligence, a decrease in its autonomy and an increase in its dependence on society, these three aspects being intimately linked. A study conducted on colonies of social insects has thus shown that brain size decreases in proportion to the size of the colonies and the degree of specialisation of individuals.[41] From a cognitive point of view, this is not surprising because intelligence develops, on the ontogenetic and phylogenetic level, when the individual is confronted with problems of various kinds: when he is not mono-tasking. Everything suggests that the same phenomenon is happening in humans. The size of the human brain has been decreasing for 10,000 years, since the Neolithic revolution. This decrease is 36 times faster than the growth it has undergone for 800,000 years. The decline in brain size since the end of the Pleistocene is explained by the fact that with the complexification of society, and especially with the lesser variety of problems we have to solve, collective intelligence is gradually replacing individual intelligence.[42] In fact, today, for many workers subjected to a repetitive and uncreative task, the number and variety of problems they have to solve are very limited, certainly more so than those faced by hunter-gatherers. I am not sure that the intellectual activity I pursue most of my time changes anything. As

[41] Riveros et al. (2012).

[42] Geary and Bailey (2009).

soon as a problem arises, I call on a specialist who will always be much more efficient than me. By doing so, our cognitive universe is dwindling and, with it, our individual intelligence. All this contributes to our loss of individual autonomy and increases our dependence on the superorganism, which ends up having a life of its own, often to the detriment of individual fulfilment. The rise and development of Artificial Intelligence could well reinforce this phenomenon by increasingly replacing the need to be individually intelligent.

5.5 The Conjunction of Ultrasociality and Individual Competition: A Monumental Obstacle to Solving the Environmental Crisis

Unlike ants, we remain individually intelligent, we continue to be obsessed with our individual success and are driven by our pursuit of pleasure and well-being, which generate an inter-individual competition absent in social insects. In their case, society has definitively crushed the individual, who has become a mere element, devoid of any autonomous life, of a superorganism with its own life and obeying objectives that completely escape the individuals themselves. Our level of alienation to the superorganism that is our society is fortunately much more limited, even if there is no doubt that the Neolithic revolution and its ultimate offspring, contemporary society, have brought us closer to this terrible condition. Social protest, non-existent in social insects, remains the most emblematic expression of a resistance that still characterises human society: an ant will never go on strike. But where the trick that evolution plays on us is dramatically powerful and constraining is when the obsession with individual success and the frenzied pursuit of pleasure and well-being cause another form of alienation necessary for the overall functioning of society, its growth, its expansionism and generally the destruction of the planet. This is clearly what the metaphor of the invisible hand means: by working for our personal interest, we contribute to economic growth, the now primary objective of any human society. But as we are eminently social beings, our personal interest, as well as our need for hedonic satisfactions, are defined only in relation to a norm that is implicitly imposed by our society. We derive pleasure from acquiring what society values and society is such that it will always value what we do not have, leading us into a consumption race in which we become the consenting slaves. We cannot escape inter-individual competition unless we accept marginalisation and exclusion that would permanently deprive us of the satisfactions we seek; which would also decrease the likelihood of passing our genes onto our descendants. For all these reasons, our individual goals converge with the goal of growth inherent in every human society. We cannot deviate from these goals unless we accept living on its margins, which is the worst punishment that can be imposed on an ultra-social animal. This probably explains why many environmentalists on paper, particularly critical and lucid towards a society whose mistakes they perceive, continue to produce and consume as society expects of them: servility as an expression of submission to the group?

By attempting to enrich ourselves individually, by trying to gain a few places on the social ladder, we collectively contribute to the enrichment of society. The inequality introduced by the Neolithic revolution, and inherent in any large, complex society capable of overproduction, is now the spur that propels us in the race for more wealth, which has become a necessary condition for economic functioning. Without this "necessary" dose of inequality, much of the energy we put into trying to enrich ourselves would disappear and inevitably lead to a decrease in economic growth. Growth is to humans what fungus is to leaf-cutter ants. It is necessary, not for the survival of individuals—Paleolithic humans coped very well with a lack of growth—but for the growth and expansionism that are the ultimate goals of society. We are individually subject to the demand for growth that feeds the superorganism we have created and which now strongly constrains our individual lifestyles. It may well be that we have given birth to a monster imposing a relationship with the world that is not beneficial to us: it is a source of inequality, it reduces our autonomy, it is a source of omnipresent, stressful and painful competition. Above all, if we fail to curb its insatiable appetite for growth and its expansionist dynamics, it will reduce our chances of survival on Earth.

The ultra-sociality of humans as expressed through parochial altruism has something very archaic about it and represents a real obstacle both to the pacification of our international relations and to the preservation of our common good, our planet. Altruism is said to be parochial because it is not universal. It is exclusively focused on our loved ones and those who are symbolically perceived as close due to arbitrary cultural similarities. It is inevitably coupled with an often aggressive desire to favour one's group at the expense of other groups. The competition between states and international military conflicts are just contemporary forms of parochial altruism as observed on a smaller scale in all primates and most social mammals. Racism, nationalism, identity withdrawal and religious intolerance are other disastrous expressions. The vaccine nationalism triggered by the Covid 19 pandemic and denounced today by the OMS is also a tragic manifestation. The consequence of our inability to extend our altruism to all of humanity is very significant in view of the environmental crisis. If a state decided to reduce its growth as part of a democratically validated political project, it would necessarily be dominated and then absorbed by those who have not broken with the economy of growth, the latter conferring on them an obvious economic and therefore geostrategic superiority. Any ecologically viable societal project, i.e. aiming at a level of growth compatible in the long term with the biophysical limits of the planet, is a dead project from the outset as long as competition continues to be exercised with states that maintain the objective of continuous growth.

The Neolithic revolution was thus the starting point of a process that leads society to grow indefinitely in a physically limited universe: a perfect antinomy. What is fascinating is that today's humans do not differ biologically and psychologically from our ancestors at the beginning of the Neolithic and even from those of the Paleolithic. Our level of intelligence has not evolved, or perhaps even regressed, our propensity for pleasure-seeking has remained intact as has our search for social status; the importance given to sex, a pleasant trap set by evolution to encourage us to pass on our genes, has not decreased; our altruism towards those close to us is manifested with as much generosity; our potential for aggression towards our competitors or individuals whose only fault is to be perceived as strangers has hardly faded, and yet, the balance has been broken: the continuation of our current trajectory is no longer sustainable. It was enough to master the cultivation of a few cereal grains for an unprecedented runaway to make what was perfectly adaptive in the Paleolithic become what urgently needs to be modified today for the survival of our species. It is as if evolution, in giving birth to the most intelligent and social species on this planet, had not foreseen that it had incidentally endowed it with the means to self-destruct and destroy the planet with it: an evolutionary success metamorphosing into an evolutionary disaster. There is nothing very surprising about this because evolution is blind and nothing excludes that it may have gone astray in terms of preserving this planet. The rest of the story will no longer depend on biological evolution. We no longer have the time. It will depend on our cultural, and therefore political, evolution. An evolution that will have a hard time because its task is immense: to counterbalance the evolutionary and biological forces that have led us to this pivotal moment in our history. As the biologist Richard Dawkins so finely expresses it,[43] the search for ecological sustainability that should prevail today is anti-Darwinian in the sense at least where ecological sustainability requires obstructing biological and psychological forces resulting from our evolution, those precisely which naturally encourage us to always grow more and to prioritise the short-term over the long-term. We must simply hope that we are not entirely the plaything of this evolution. Our large brain should be capable of developing new values and new objectives compatible with the preservation of the environment. This is what it has begun to do for a few decades now but has not yet come to fruition, far from it.

References

Biraben, J.-N. (2003). The rising numbers of humankind. *Population & Societies, 394*, 1–4.

Bocquet-Appel, J.-P. (2011). When the world's population took off: The springboard of the Neolithic demographic transition. *Science, 333*, 560–561.

Boehm, C. (1999). *Hierarchy in the forest*. Harvard University Press.

Boehm, C. (2013). The biocultural evolution of conflict resolution between groups. In D. P. Fry (Ed.), *War, peace, and human nature* (pp. 315–340). Oxford University Press.

Bowles, S. (2011). Cultivation of cereals by the first farmers was not more productive than foraging. *Proceedings of the National Academy of Science, 108*(12), 4760–4765.

Byrne, R. W., & Whiten, A. (1988). *Machiavellian intelligence: Social expertise and the evolution of intellect in monkeys, apes, and humans*. Clarendon Press/Oxford University Press.

Chagnon, N. (1977). *Yanomamo: The fierce people* (2nd ed.). Holt.

[43] Dawkins (2001).

Cohen, M., & Crane-Kramer, G. (2007). *Ancient health: Skeletal indicators of agricultural and economic intensification*. University Press of Florida; Patricia Lambert. Health.
Dawkins, R. (2001). *Sustainability doesn't come naturally: A Darwinian perspective on values*. The Environment Foundation Inaugural Lecture.
Derex, M., Beugin, M. P., & Godelle, B. (2013). Experimental evidence for the influence of group size on cultural complexity. *Nature, 503*, 389–391.
Diamond, J. (1978). The Tasmanians—The longest isolation, the simplest technology. *Nature, 273*, 185–186.
Divale, W. T., & Harris, M. (1976). Population, warfare, and the male supremacist complex. *American Anthropologist, 78*, 521–538.
Dortier, J.-F. (2015). *Revolution in our origins*, under the direction of Jean-François Dortier. Editions, Human Sciences.
Dunbar, R., & Shultz, S. (2007). Understanding primate brain evolution. *Philosophical Transactions of the Royal Society B, 362*, 649–658.
Fittkau, E., & Klinge, H. (1973). On biomass and tropic structure of the central Amazonian rain forest ecosystem. *Biotropica, 5*, 2–14.
Gat, A. (2006). *War in human civilisation*. Oxford University Press.
Geary, D., & Bailey, D. (2009). Hominid brain evolution: Testing climatic, ecological, and social competition models. *Human Nature, 20*, 265–279.
Gowdy, J. (Ed.). (1998). *Limited wants, unlimited means: A reader on hunter-gatherer economics and the environment*. Island Press.
Gowdy, J., & Krall, L. (2014). The transition to agriculture and the evolution of human ultrasociality. *Journal of Bioeconomics, 16*(2), 179–202.
Hayden, B. (2008). *Man and inequality. The invention of hierarchy in prehistory*. CNRS Editions.
Henrich, J. (2004). Cultural group selection, coevolutionary processes and large-scale cooperation. *Journal of Economic Behaviour and Organisation, 53*(1), 3–35.
Hilt, B. M., & Formicola, V. (2008). Hunters of the ice age: The biology of upper paleolithic people. *Yearbook of Physical Anthropology, 1*, 70–99.
John Gowdy, J., & Baveye, P. (2019). An evolutionary perspective on industrial and sustainable agriculture. In G. Lemaire (Ed.), *Agro-ecosystem diversity: Reconciling contemporary agriculture and environment quality*. Elsevier.
Jones, S. (1993). *The language of the genes*. Anchor Books.
Kaplan, H., Lancaster, K., & Hurtado, M. (2000). A theory of human life history evolution: Diet, intelligence, and longevity. *Evolutionary Anthropology, 9*, 156–185.
Knauft, D. (1994). Culture and cooperation in human evolution. In L. Sponsel & T. Gregor (Eds.), *The anthropology of peace and nonviolence* (pp. 37–67). Rienner.
Larsen, C. S. (2006). The agricultural revolution as environmental catastrophe: Implications for health and lifestyles in the Holocene. *Quaternary International, 150*, 12–20.
Majolo, B. (2019). Warfare in an evolutionary perspective. *Evolutionary Anthropology, 28*, 321–331.
Pathou-Mathis, M. (1993). Paris, Odile Jacob.
Ponting, C. (2007). *A new green history of the world*. Penguin Books.
Richerson, P. J., & Boyd, R. (2005). *Not by genes alone: How culture transformed human evolution*. University of Chicago Press.
Riveros, A., Seid, M., & Wcislo, W. (2012). Evolution of brain size in class-based societies of fungus-growing ants (Attini). *Animal Behaviour, 83*, 1043–1049.
Sahlins, M. (1968). *The original affluent society*. (Reprinted in Gowdy, 1998) (pp. 5–41). Routledge.
Sandom, C., Faurby, S., Sandel, B., & Svenning, J.-C. (2014). Global late Quaternary megafauna extinctions linked to humans, not climate change. *Proceedings of the Royal Society B, 281*, 20133254. https://doi.org/10.1098/rspb.2013.3254
Smith, E. A., & Wishnie, M. (2000). Conservation and subsistence in small-scale societies. *Annual Review of Anthropology, 29*, 493–524.

Tajfel, H., Billig, M. G., Bundy, R. P., & Flament, C. (1971). Social categorisation and intergroup behaviour. *European Journal of Social Psychology, 1*, 149–178.

Tomasello, M. (2009). *Why we cooperate*. MIT Press.

Turchin, P., Currie, T. E., Turner, E. A. L., & Gavrilets, S. (2013). War, space, and the evolution of Old World complex societies. *Proceedings of the National Academy of Sciences USA, 110*, 16384–16389.

Wilson, E. O. (2012). *The social conquest of the earth*. Liveright. W.W. Norton.

Wilson, D. S., John, M., & Gowdy, J. (2013). Evolution as a general theoretical framework for economics and public policy. *Journal of Economic Behavior & Organization, 90*, S3–S10.

Wrangham, R. W. (2018). Two types of aggression in human evolution. *Proceedings of the National Academy of Sciences USA, 115*, 245–253.

Wright, R. (2004). *A short history of progress*. House of Anansi Press.

Zvelebil, M., & Rowley-Conwy, P. (1986). Foragers and farmers in Atlantic Europe. In M. Zvelebil (Ed.), *Hunters in transition* (pp. 67–96). Cambridge University Press.

Chapter 6
The Boredom and Horror of Meaninglessness: The Ultimate Obstacle

Abstract Humans are not perfectly equipped to handle boredom and the absence of desire. When boredom and lack of desire take hold, it results in a state of psychological suffering that can lead to depression and acute anxiety. There are various ways to manage boredom and lack of desire, and the emergence of post-Paleolithic societies has played a fundamental role in this regard. These societies, by making inequality a norm, have made possible the ever-renewable desire for consumption and personal enrichment. Capitalism is very effective from this point of view, hence the difficulty we have in escaping it, even though we know its destructive dimension at the environmental level.

Man is a decidedly original animal. He is incapable of surviving the disappearance of all forms of objective; the idleness that results sends him explicitly or implicitly back to the horror of meaninglessness: the first and most commonplace of metaphysical anxieties. This type of experience can take various forms. It can be the simple feeling of boredom as can be felt by the idle child and temporarily deprived of his friends, his video games or any other form of occupation. It can also be the adult who does not know what to do with his free time at the weekend, the unemployed or the retiree finally freed from the grip of his work and confronted with the torment of time unconstrained. It can still be a more conceptual question, of a philosophical or metaphysical nature, trying to identify a meaning to our universe and our place in it. Never as much as when we are freely left to ourselves, this type of questioning becomes pervasive. Although these mental states may seem different, they remain psychologically determined by the experience, unbearable for most humans, of emptiness. An emptiness that can quickly turn into anxiety or depression. Men are decidedly very paradoxical. They often aspire to holidays, free time, retirement, but what wouldn't they do to be busy no matter what it costs as soon as they can no longer conceive of any objective that can fill the void that inevitably

T. Ripoll, *Human Psychology and the Destruction of the Planet*, Anthropocene – Humanities and Social Sciences, https://doi.org/10.1007/978-3-032-02080-2_6

generates idleness. The level of anxiety generated by boredom varies of course from one individual to another. Some will go as far as to immolate themselves on the altar of work to escape the painful grip of emptiness while complaining about the pressure they are under: simple socially valued coquetry intended to hide the fact that they work to avoid confronting the terrifying monstrosity of boredom. Others will find a number of small adjustments in their life to escape it, such as these derisory and small diversions or occupations without a future that have no other interests than to "kill time" to resume a popular expression of extreme lucidity.[1] A few rare others seem better equipped, either because they are capable like these satiated felines of daydreaming for hours on end in their deck chair, or because they are used to meditation techniques that have been very fashionable for a few years. Generally speaking, I am quite convinced that a large part of the energy that we deploy, when it is not used to fulfil objectives that are imposed on us by everyday life, is primarily intended to build ramparts against the anxiety generated by boredom and emptiness. From an evolutionary point of view, this is not very surprising. What adaptive advantage would there be for our brains to have evolved so that they can take pleasure in inaction and confrontation with boredom? Probably none, especially in the natural universe in which we lived just a few tens of thousands of years ago. If our ancestors of the genus *Homo* had had the ability to take pleasure in doing nothing, if they had been content with total mental and physical inactivity once their basic needs were met, it is likely that *Homo Sapiens*, this perpetual excited and anxious being, would never have seen the day. We have been evolutionarily formatted to interact actively with reality, to try to appropriate it, to shape it, to transform it to our advantage. This is why humans can hardly bear to be confronted with the absence of objectives, an absence too inclined to lead to a meaninglessness that only the brains of *Sapiens* have the painful privilege of perceiving. When this happens, it is necessary to quickly find the parade. This usually consists in inventing objectives that can then be substitutes for the current objectives that were ours in the past. The addiction to video games is a remarkable illustration of this and it is not surprising that it is so common among teenagers or young adults who can no longer identify satisfying objectives in their real life. The video game is the very example of a substitute, not very demanding on the cognitive level, for the objectives that men usually follow when they manage to effectively face the real world. In most cases, the objectives of video games are: conquest, destruction of adversaries, increase of wealth, construction, survival in a dangerous universe and, of course, sexual conquest: all the objectives that humans have been following for a long time once their food needs are met. But the incredible success of video games, and therefore the addiction they generate, is largely explained by the fact that the objectives that the player must achieve are systematically renewed. As soon as one objective is achieved, another one appears. Profile and so on to infinity so as not to give any chance for emptiness to surface. Video games have thus managed to do much better what would be the Holy Grail of consumer society: renewing desire as soon as it has been satisfied.

[1] Bronner (2021) considers that our addiction to the Internet is partly linked to the torture that boredom represents for our mind.

6.1 Objectives and Desires: The Best Remedy for Boredom

Let's imagine the following thought experiment that I will call "the torment of the satiated": a radical form of Aladdin's magic lamp. Let's imagine that a man is given the ability to satisfy every single one of his desires at any given moment. The moment he starts dreaming of living in a magnificent mansion, he finds himself transported to the mansion of his dreams. The moment he would like to visit the lagoons of Polynesia, he finds himself with his feet in transparent water beautifully coloured with coral fish. Here he is, excited at the thought of a wonderful young woman lavishing him with all the naughty treats he had never before imagined, and she presents herself to him in the simplest attire, welcoming and smiling. Like any good supporter, he hopes that his favourite football team will win the World Cup and, immediately, three goals are concede by the opposing team. In short, any desire, of whatever nature, would be satisfied the moment it is conceived without even a small parcel of time being able to intervene between the desire and its satisfaction. Unless one lacks empathy or is excessively subjected to frustration in one's own life, it is not very difficult to conceive of the absolute suffering that would afflict this man. What would become of a life that would not even allow us to genuinely aspire to anything because it would be so easy to obtain it immediately? A life where the words "waiting, hope, objectives, wishes..." would no longer have a place, leaving room for a generalised emptiness, boring and despairing. Undoubtedly, the only solution to escape the torment of the satiated would be to ask for the omnipotence to cease, so that this man could desire again and especially desire without being certain of obtaining the expected satisfaction. No longer feeling the resistance of reality, because it would be too generous, is the best way to abolish desire and to raise our level of anxiety related to the absence of meaning in life. The Buddhist solution is quite extraordinary from this point of view since it consists, *a contrario* of what this thought experiment suggests, in working to directly make desire disappear in order to escape the suffering of frustration inherent in life. On paper, such a theoretical objective is remarkable. It is even fascinating that men have been able to conceptualise such an unnatural relationship to the world and so magnificently detached from our coarse and real animality. However, I doubt that such an objective can be applied to humans simply because it amounts to denying what constitutes our being as a biological organism animated by drives that evolution has inscribed in us so powerfully. In reality, as everyone knows, the end of desire is also the end of life and it is not at all obvious that one can escape one without ending the other. The end of desire is probably the last of the mental states that humans can bear. It is not impossible that it is even more unbearable than radical frustration as the absence of desire is psychologically unbearable. Keynes had analysed it very well, noting that once our basic needs are met and our desires satisfied, "for the average individual, devoid of special talents, it is a real problem to find something to do". In a certain way, I think that if the ability to endure boredom is not adaptive, the ability to desire is totally so. Evolution has done things well: it has made us beings of desire incapable of surviving boredom, which is why the immediate satisfaction of our desires is the worst thing that can happen to us.

It is not very difficult to understand that our propensity for desire in all its forms has been the driving force of human life and of course of biological evolution as well as cultural and technical evolution. I have often wondered about the underlying dissatisfaction that seems to be present in most humans. Not necessarily a dissatisfaction that causes suffering, but at the very least a dissatisfaction that makes us move forward, evolve, progress when we manage to implement the necessary means to overcome it. This predisposition to dissatisfaction, coupled with the inability to resist boredom, fuels human hyperactivity, their propensity to grow, but also to understand, to discover, to search, and therefore not to be fully satisfied with their state, even if it is objectively very pleasant and satisfying.

In the thought experiment of the torment of the satiated, the last satisfied desire I mentioned is that of the victory of his favourite football team. It is very easy to imagine that the spectacle of a football match would be greatly diminished if the outcome was always predictable. Obviously, it is not so much winning that counts, but the fact of winning when the outcome is uncertain, the fact also of winning in adversity. Nothing is as pleasant as what one has obtained through hard struggle. It is from this uncertainty and adversity that desire is born and survives, as well as the real pleasure or happiness of being able to satisfy it. We have a vital need for this delicious discomfort.[2] The consequence of all this is that the interindividual competition at the heart of our liberal societies, the major disadvantages of which have been clearly identified, is probably also a necessary ingredient for human psychological balance. We can indirectly measure this through the almost universal enthusiasm that humans accord to sports competition. What is at stake when we watch a sports competition? Many things, but above all the pleasure, by proxy, of winning and getting the upper hand over one's opponent. This is so powerful that the supporters of a player or a team see their testosterone levels increase in case of victory, directly contributing to increase their sexual desire. But this form of rather archaic excitement, probably a ritualised form of the intraspecific competition that all animals of a given species engage in to assert themselves in their group, only makes sense if the outcome cannot be known in advance. The very fact of being able to lose conditions the possibility of being happy to win. What would become of the Tour de France if it was simply about watching cyclists calmly climb the slopes of the Alpine passes? Whether they ride at 10, 20 or 50 km/h would hardly matter in view of our passion for victory and defeat. Would we take the same pleasure in watching Nadal and Federer play tennis if it was no longer possible to count the points and experience the adrenaline surge caused by the match point? In the fight against boredom and in the perspective of increasing the level of uncertainty of the outcome of a competition, the most extreme solution would be to put one's life at stake in a perfectly gratuitous way. From an evolutionary point of view, this would be perfectly maladaptive as dying for nothing is of little interest in terms of species survival. Yet,

[2] Belk et al. (2003).

this is a practice as old as humanity that has been institutionalised in certain risky sports, notably in bullfighting, to the point that their heroes ultimately derive real benefits from it. But other unknowns embark on totally gratuitous and often deadly challenges, leading them to confront the verticality of cliffs with bare hands or the monstrous waves of the oceans on board frail skiffs. Faced with the horror and imminence of death, boredom and meaninglessness are definitively defeated by this type of high-risk activity. Only the human species is capable of this feat as it is important for it to defy meaninglessness and boredom, against which our human brains are totally helpless. Rather die than be bored, such could be and sometimes is the logic that guides our behaviours.

6.2 Another Brake on the Ecological Revolution

Man is generally hyperactive and remains so even when his vital needs are totally satisfied. This hyperactivity, probably inscribed in our genes, is what is at the origin of our evolutionary dynamic, both cultural and biological. If the Neolithic revolution has imposed itself, it is also for this reason. It allowed men to break away from the soft and comfortable humdrum of the hunter-gatherer lifestyle, not that life in the Paleolithic was comfortable and free of all danger, but rather that its extreme stability, its perennial character and its intrinsic immobility must have paled in comparison to the attractiveness of a Neolithic culture already oriented towards growth, innovation, progress, discovery and personal enrichment. The average human is incapable of resisting the anxiety generated by boredom and the Neolithic revolution, because it invented the dynamic of continuous growth and the possibility of personal enrichment, constituted the most powerful antidote to the risk of vacuity. Its psychological attractiveness also lies in this. The waves of European migrants who joined the USA in the last centuries, those of Africans who are flooding into Europe today, are not only the result of their modest or sometimes miserable living conditions in their country. They also result from the attractiveness, sometimes totally fantasised and dangerous, of foreign cultures where everything seems possible, notably the opportunity to enrich oneself, to evolve and to be animated by new desires that success may perhaps satisfy.

In reality, inequality and the possibility of ever-increasing consumption probably constitute both the most attractive and the most contested characteristics of our society. Instituted inequality as a new norm of social life, and constantly renewed consumption, are what human has invented best to escape boredom and maintain his need for desire at a high level. As they also constitute the main drivers of economic growth, we are faced with a conjunction of forces from which it will be very difficult to free ourselves. Inequality is often fought and I am one of those who think it is extremely dangerous and ethically unacceptable, especially when it is pushed to the point of no longer allowing the poorest to even hope to escape their condition. However, history shows that the ideal of social justice and equality is often bypassed by those who have defended it fiercely and often authentically: is this a contingent phenomenon and does it reveal a powerful component of the human psyche? How many leaders of communist parties have become enormously rich after successfully leading their revolution? Equality yes… but for others. As for the citizens of communist countries, it is not in itself the fundamental inequality of Western countries that so powerfully seduced them, but its positive counterpart: the possibility of becoming richer than others. Equality has something boring about it when inequality only reveals, for the optimist, the potentiality of personal success. Men are so made that they prefer to take the risk of losing than to live the boredom of a life where it is possible neither to lose nor to win. Experimental psychology has confirmed the power of this sad passion. If subjects are asked whether they prefer to live in a society that allows them to earn more but to earn like others, or to live in a society where they will earn less but can earn more than others, the majority will choose the second option. Of course, the obsession with social status partly explains this result, but being deprived of the opportunity to earn more than others appears as an unbearable deprivation of freedom: that of being deprived of the risk of defeat and the chance of victory. Beyond, and more deeply, this deprivation of freedom threatens to bore human who could neither succeed nor fail due to social norms that would prevent social differentiation. Such is the terrible attractiveness of capitalism, such was also the power of the Neolithic revolution.

From this point of view, the failure of communism, as well as its authoritarian and dictatorial component, is probably not an accident of history. Theoretically, egalitarianism and the refusal of private property can be perfectly justified. There is probably a form of unquestionable ethical rationality in this objective. But this rationality comes into direct conflict with the archaism of psychological forces that can be contested but which work underground in all humans. Forces resulting from our evolution which were exacerbated by the Neolithic revolution. These forces are such that they express themselves in one way or another as soon as a space of freedom presents itself. This is the reason why communism struggles to do without authoritarianism when capitalism perfectly accommodates, even claims and imposes, a form of freedom that cannot be contested. A freedom that serves as an alibi when society tries to legislate to protect the citizen-consumer from the wanderings of harmful consumerism. On the contrary, communism does not get along with desire because this latter is the engine of interindividual competition, the very one which naturally leads to the social inequality it fights. Conversely, capitalism feeds on desire. It exacerbates it through advertising because it is from constantly renewed desire that it draws its economic power and from it again that it obtains the fuel necessary for growth and therefore for its survival. I absolutely do not want to praise here a model of society and a liberal economy which I am convinced will be fatal to men in the long term. I simply underline the fact that this model is "natural" to quote the words of Gowdy and Krall[3] in the sense that it is a mode of production and consumption that draws its strength and its universality from a well-identified evolutionary dynamic. Overturning it, or finding a viable substitute for it, will be

[3] Gowdy and Krall (2013).

absolutely necessary but can only succeed by becoming aware of the deep reasons that led it to colonise the planet in a few millennia. Believing that it only results from the will of the wealthy and powerful to maintain it in order to preserve their advantages is just part of the explanation, probably not the most decisive, but the one that is most convenient to identify the culprits of such a situation.

As for consumption, if it is explained by a multitude of previously exposed psychological factors, it has this incomparable and sneaky privilege of constituting a remedy all the more effective against boredom as it does not actually lead to any real lasting satisfaction: it is just good at generating desire without ever managing to satisfy it. It thus constitutes the perfect mirage that men need to kill boredom and escape the abolition of desire, and therefore the fuel that our society needs to continue to grow.

If this interweaving of factors leading to the destruction of the planet were not the simple result of our evolutionary trajectory, one could see it as the perfect coup of a very sophisticated Machiavellian plan. In reality, there is no plan. Nature knows how to produce extraordinarily complex things without being guided by any intention. It will be up to us, as far as this is possible, to invent a society capable of mastering the powerful forces that are leading us at high speed to the destruction of this tiny planet. Finding ways to reduce the level of social inequality and the overall level of consumption in the perspective of saving the planet are two objectives that cannot be dissociated because they are biologically, psychologically, ideologically and politically linked.[4] But to believe that it is possible to achieve them once power escapes from the powerful of this world is very naive. There will probably be new powerful to replace the fallen powerful. The psychological determinisms that have been at work until now will remain intact. We do not erase millions of years of evolution with a wave of the hand. Denying them or trying to control them authoritatively is probably not the right solution. The path will therefore be narrow.

References

Belk, R. W., Lizger, G., & Askegaard, S. (2003). The fire of desire: A multisited inquiry into consumer passion. *Journal of Consumer Research, 30*, 326–351.

Bronner, G. (2021). *L'Apocalypse cognitive*. Paris, PUF.

Feygina, I. (2013). Social justice and the human–environment relationship: Common systemic, ideological, and psychological roots and processes. *Social Justice Research, 26*, 363–381.

Gowdy, J., & Krall, L. (2013). The ultrasocial origins of the Anthropocene. *Ecological Economics, 95*, 137–147.

[4] Feygina (2013).

Chapter 7
The Pursuit of Personal Interest and the Tragedy of the Commons: The Strength of the System

Abstract In a now-famous article, Hardin developed the foundational concept of the "tragedy of the commons". Hardin thus shows that if individuals managing a common good prioritize their particular interest, it necessarily results in the destruction of the common good. This thesis applies without difficulty to the management of our common good: our planet. Various authors have challenged Hardin's analysis. However, I show here that the characteristics of our societies only reinforce Hardin's analysis. Unless we profoundly change our economic paradigm, we cannot avoid the tragedy of the commons.

Let's recap. Like all animal species, human is subject to intraspecific competition provoked by the race for social status: the search for the hierarchical position most conducive to the transmission of his genes through his offspring. This race is today one of the engines of economic activity and at the origin of a rivalry that manifests itself at all levels of society: at the individual level, at the level of companies and at the level of states. Until the Neolithic, this intraspecific competition was expressed in a more symbolic than material way because a fundamental egalitarianism excluded any attempt at personal material enrichment. The Neolithic "liberated" human from this constraint and instituted inequality as a new social norm. With it, arose the joint possibility of producing and consuming more than necessary. The cultural bases of infinite economic growth were thus laid. Moreover, human finds in this new relationship to the world a powerful remedy for emptiness because the very fact that it has become possible to enrich oneself always more creates a potentiality of objectives that allow to indefinitely fuel desire and frustration. Men have thus invented a system that guarantees that opulence is always aimed for but never achieved.

The overall picture that emerges from all this is that of a human being obsessed with his own interest and driven by fundamental selfishness. Both would constitute

T. Ripoll, *Human Psychology and the Destruction of the Planet*, Anthropocene – Humanities and Social Sciences, https://doi.org/10.1007/978-3-032-02080-2_7

the major psychological obstacles to the coordination and cooperation necessary for the preservation of the common good that is our planet. This is at least what ecologist Garrett Hardin wanted to express in a famous 1968 article[1] "The tragedy of the commons", which can be said to have initiated ecological economics. The tragedy of the commons is a thought experiment featuring farmers who graze their cows on a shared, freely accessible piece of land. If the pasture is of limited size and each farmer is content with the cows he owns, the overall state of the system is stable. It remains in balance as long as the amount of grass consumed by the cows does not exceed what the pasture can produce. Such a balance requires that growth, in terms of cows raised, becomes null as soon as the amount of grass consumed reaches or exceeds the productivity of the pasture surface shared by the farmers. But what happens if one of these farmers, wishing to enrich himself further, decides to acquire an additional cow and add it to those already grazing? Once the cow is fattened, he can resell it and obtain a profit from which he will be the sole beneficiary. The economic utility of such a decision is obvious: the farmer has made a rational decision economically speaking since he is simply acting with the aim of further enriching himself. The problem is that by acquiring an additional cow he will contribute to exploiting the common good: the pasture. In fact, the cost of this additional exploitation will be spread over all the farmers since the grass consumed by this additional cow will no longer be available for the other cows. If initially there are 10 farmers who had an equivalent herd, the investing farmer will in reality only pay one tenth of what was used to raise his own cow, the remaining nine tenths being involuntarily borne by the other nine farmers. The farmer will incidentally make the community pay for the cost of raising a cow of which he is the sole owner and from which he will be the sole beneficiary. The risk of such a situation is obvious. If each owner behaves in this way—why wouldn't they?—that is, by only taking into account their own interest and neglecting the collective interest, overgrazing will be such that the resource will soon be exhausted, directly endangering the continuation of their activity. This thought experiment is just a parable of the overexploitation of the living by companies or states obsessed with their economic growth. In a certain way, the environmental crisis is only the consequence of the primacy of personal interest over collective interest. The common good being limited, if everyone selfishly tries to enrich themselves through the resources it contains, the tragedy is inevitable. The escalation in the race for personal enrichment inevitably leads to the collapse of the system, hence the term tragedy adopted by Hardin, who considers that, without some form of regulation, human psychology systematically turned towards personal interest necessarily leads to collapse.

Hardin envisaged two radically different types of solutions to the tragedy of the commons: private property or the regulatory and coercive power of the state. A large part of the success of his work was probably due to the fact that ecologists, neoliberals and communists could all claim him. Ecologists, because obviously the thought experiment shows that infinite exploitation of our environment is not possible;

[1] Hardin (1968).

communists, because unregulated individual interest by a state force leads to the disaster of the common good; and liberals, because private property would be the best guarantee of preserving a resource.

7.1 Was Hardin Right?

Hardin was an ecologist heavily influenced by the work of Darwin and Malthus. He can be considered a neo-Malthusian. He believed that any human population naturally and excessively tends to grow. If demography is not regulated, overpopulation would inevitably lead to the exhaustion of resources, famine, violence and, in the case of humans, the collapse of society, he thought. With Jarett Diamond,[2] Hardin is probably one of the fathers of contemporary collapsology. Towards the end of his life, Hardin defended a radical position on the need to reduce the human population, associated with a form of authoritarianism and xenophobia that is very problematic. The Hardin couple committed suicide the day after their wedding anniversary, not without having had four children, which is a lot for a couple obsessed with the necessary reduction of world demography.

Hardin had in his sights a naive conception according to which natural resources are inexhaustible. It is paradoxical that the affirmation of the inexhaustible nature of natural resources was clearly formulated in the seventeenth century, at the very moment when men, having conquered almost the entire planet, could finally perceive its contour and question the limits of its exploitation. Perhaps it was the fear of being subject to these limits which curiously led to the absurd assertion of a lack of limits: is not the best way to deny a reality to assert exactly its opposite? In 1625, a Dutch jurist and Protestant, Hugo de Groot, wrote "On the Freedom of the Seas", a work in which he claimed the possibility of freely exploiting the resources of the sea, a work which is partly at the origin of contemporary international maritime law. The position he defended is liberal in economic terms and perfectly utopian ecologically since he considered that the ocean was too immense for its exploitation by all the states of the planet could one day pose a problem. De Groot's conclusion is clear: let men freely and as much as possible exploit the riches of the ocean. We can today smile at so much credulity and ecological incompetence. Yet, many political leaders and intellectuals continue to succumb to such a crude utopia. The cornucopian version of sustainable green growth is but a sophisticated form of de Groot's position. Hardin also had in his sights the liberalism of Adam Smith as it was powerfully developed in "The Wealth of Nations", the founding work of modern capitalism and written 150 years after de Groot's work. From Smith's perspective, when economic actors—the breeders in Hardin's parable—operate freely and work for their personal interest, the invisible hand will in reality promote the common interest. By allowing the selfish interests of economic actors to express themselves without constraint, it will result in a global enrichment that will serve the community.

[2] Diamond (2005).

The conjunction of a belief in the infinite nature of natural resources and that of the power of the invisible hand to always converge towards an economic optimum is, for Hardin, extremely dangerous for the future of the human species. This is precisely the objective of his thought experiment to reveal the disastrous consequences.

Like any analogy, the thought experiment proposed by Hardin has its own limits, but it must be acknowledged that the facts have largely proved him right. If we take the paradigmatic example of the fishery resources of the oceans and the whaling industry, the facts are clear. All populations of commercially exploited predator fish have reached a critical state.[3] Some are simply no longer profitable given their rarity and others are doomed to disappear forever because, below a certain threshold, the survival of the species becomes impossible even when it is no longer exploited. The case of whaling is even more representative. Without the advent of oil which replaced the oil extracted from large cetaceans, most of these would have disappeared in the twentieth century even though large-scale hunting only lasted for two centuries. The grey whale has nevertheless definitively disappeared from the Atlantic Ocean and only 1% of the original population of blue whales currently survives even though the hunting of this whale was banned by the International Whaling Commission in the 1960s. The collapse of great historical societies, as exposed by Jared Diamond in his bestseller "Collapse", seems to support Hardin's thesis. This was probably the case of the Sumerians, 4000 years BC, who overexploited their arable land, as was the case with the Mayans in their uncontrolled management of their natural resources. The fashion for collapsology is undoubtedly excessive, reviving the great biblical myth of the Apocalypse which some revel in a possible disastrous outcome. Nevertheless, whatever the excesses of collapsology and the too quick and sometimes simplistic explanations of the decline of great civilisations, one cannot deny the fact that great cultures have suffered considerably from a reckless management of their resources.[4] Today the unbridled exploitation of fossil fuels, mineral resources or the management of CO_2 in the atmosphere are well-known examples of common goods that we fail to manage rationally.

Overall, however, the tragedy of the commons rests on two premises, one of which is indisputable while the other is not. It is indisputable that the resources of the Earth are limited and that it is not possible to exploit them ever more. It is debatable that particular interests always take precedence over the general interest. The question is therefore that of the relative strengths of altruism and selfishness in human psychology.

[3] Hutchings and Reynolds (2004).

[4] Downey et al. (2016).

7.2 The Survival of the Planet Depends on the Domination of Altruism over Selfishness

One of the master ideas of Darwinism is that each individual is driven by the will to transmit his genes to his descendants: our obsession with sexuality and love is the most visible consequence. The more an individual is adapted to his environment, the more he has chances of surviving and the greater is the probability of having a descendant, therefore of transmitting his genes. This elementary process has led species to evolve optimally in relation to the constraints of their environment. In other words, intraspecific competition is a blind process leading to an adaptive optimum for the species and no longer just for the individual. In the framework of Darwinism, selfishness is therefore a fundamental property of the individual, the one that gives him the most reproductive possibilities. Altruism is much more problematic conceptually speaking because it is not so easy to identify its function in the context of intraspecific competition. What advantage could draw an altruistic individual to behaviour that promotes the survival of others, and sometimes, of their competitors? By definition, the altruistic act diminishes the selective value of the individual who exhibits such behaviour. Altruism does not seem evolutionarily stable as the most selfish individuals seem better equipped to emerge victorious from the competition for life. If this is the case, altruism should have gradually disappeared from the behavioural repertoire of all animals: the adaptive value of altruism is therefore not self-evident.[5]

Yet, ethologists and psychologists alike have highlighted the existence of altruistic behaviours in a great many living beings, from the most elementary bacteria to non-human primates and humans, of course.[6] Many animals, birds and mammals, emit warning signals to individuals in their group when a predator is perceived. This presents a great adaptive advantage for the group but not for the altruistic emitter who makes themselves more visible to the predator by signalling acoustically. Some birds do not hesitate to expose themselves to a predator to attract its attention and allow the survival of more fragile individuals. A remarkable case of altruism is observed in vampires. When an individual has taken blood from an animal to feed, it is common for them to donate part of their loot to another hungry individual in the colony.[7] Altruism can go very far in social insects, some individuals can in a way give their lives when they carry a disease transmissible to the colony, not to mention a form of biological altruism leading individuals, the workers or soldiers, to renounce their reproductive ability to devote themselves exclusively to the work necessary for the smooth functioning of the colony. It is not uncommon for chimpanzees to adopt orphaned newborns and care for them as if they were their own child, just as mutual aid and collaboration are extremely common in many of their daily activities. Cases of interspecific adoption are not rare, especially in captivity,

---

[5] Boehm (1997, 2008); Van Veelen and Hopfensitz (2007).

[6] de Waal (2008).

[7] Denault and McFarlane (1995).

predators like panthers can adopt natural prey such as gazelles. These examples among thousands of others are not intended to convince us of the existence of a "Care Bears" universe. They simply reveal that altruism, in very different forms,[8] is very common in living beings and has undoubtedly played an important role in the survival of species. More generally, contemporary ethologists admit that most mammals have a high sense of ethics and show a rejection of injustice. All these behaviours present no apparent adaptive advantage for the altruistic individual, but are very favourable to the survival of the group, and therefore of the species. In a way, the most social species, those that adopt the maximum of altruistic and cooperative behaviours are also those that dominate the planet: this is of course the case of social insects.[9] It is especially the case of humans. We should not infer from all this that the altruistic act is intentional and fully conscious; this goes without saying in social insects or "altruistic" bacteria. Nor should we infer that the altruistic act is devoid of any search for personal interest: one can give to better receive, altruism then being only a form of disguised selfishness. But it doesn't really matter: altruism is as basic a component of living beings as selfishness can be.

The paradox that altruism represents in a fairly elementary Darwinian framework has given rise to a multitude of theories and sophisticated mathematical models. All converge at least towards the recognition of the fact that altruism, whatever its forms, has a real adaptive value, exclusively selfish behaviours being likely to lead to evolutionary dead ends. The case of cancer cells growing indefinitely is symptomatic of this type of failure, the "selfish" cells of cancer leading to the death of the patient and therefore to their own death: a biological form of a tragedy of a common which, in this case, is none other than the human body.

Humans are therefore no more fundamentally selfish than they are altruistic and the outpourings of free generosity that natural disasters elicit are a remarkable illustration of this, especially since in this case, altruism is directed towards victims who are perfect strangers. From this point of view, the neoclassical conception of the human being as *Homo economicus,* that is, of a human being whose fundamental rationality consists in constantly trying to maximise his profit, has little psychological plausibility. The hypothesis of the invisible hand finds its limits here because the systematic, blind and exclusive search for personal interest is often no more to the advantage of the individual than to the advantage of the species or society. Above all, if humans, probably even more than all other higher animal species, are capable of altruism then the tragedy of the commons is no longer inevitable. It might be possible that because of their altruism leads people to sacrifice a part of their personal interest to preserve the common good. The term 'tragedy' chosen by Hardin to express the inevitability of the phenomenon would be excessive in this case. We would have to face the problem of managing the commons and we would not be the victims of a natural selfishness that would determine the course of the world until the final collapse. This is precisely the sense of the criticism that the Nobel laureate

[8] Alberto and Ravat (2011).

[9] Gowdy and Krall (2016).

in economics, Elinor Ostrom, addressed to Hardin. For Ostrom,[10] Hardin is caricatural. He implies that the breeders are strictly determined by their natural selfishness without being able to agree, exchange and communicate, in order to reach a reasonable agreement to avoid disaster and preserve their common good.

7.3 Conditions not Conducive to the Transition from Selfish Behaviour to Altruistic Behaviour

If altruism is infinitely more widespread than initially thought, some conditions are more favourable than others. The question arises as to whether the need to preserve the common good that is our planet is likely to arouse altruism and cooperation without which we are heading for the disaster predicted by Hardin: this is the heart of the problem.

To take one of the most obvious threats, let's take the example of global warming caused by the production of CO_2 in the atmosphere. In this specific case, the common good that we need to preserve is the atmosphere. Human psychology is such that we will only agree to move from selfish behaviour leading to the destruction of the common good to altruistic behaviour allowing its preservation, if two fundamental conditions are met. The first is to be able to very concretely represent the tragedy that is unfolding before us. The second is to be convinced that a change in one's own behaviour can have an effect on the tragedy.

Regarding the first condition, the situation we are facing is much more formidable than the one Hardin describes in his parable. The pasture is a very concrete and human-scale material reality that it is possible to temporally and physically grasp its deterioration. Our breeders could very easily visualise the state of their pasture and they could just as concretely perceive the impact of their cows on the state of the common good. But can we in the same way represent the destruction of fossil energy or, even worse, the concomitant increase of CO_2 in the atmosphere? No, in reality, because humans have great cognitive difficulties in correctly evaluating such gigantic quantities or, even worse, statistics concerning them. No one experiences the increase of CO_2 in the atmosphere because no one perceives the CO_2. Our cognitive limitations are such that scientific and quantitative information on the degradation of such a common good is unlikely to have a real impact on our behaviour. Barely perceiving the devastating impact of our way of managing this common good, we continue to exploit it and will probably continue to do so until the consequences directly and concretely impact our way of life. The human brain needs concrete stimuli to be able to initiate a real behavioural change. This is how it has been observed that a single heat wave, which in itself means nothing in relation to global warming, has more impact on individual behaviour than all the statistics and scientific data revealing the seriousness of the phenomenon. Of course, the level of

[10] Ostrom (1990).

scientific education plays a decisive role here as it allows us to more concretely grasp dramatic developments that are non-significant and too abstract for a majority of humans not used to analysing quantitative data. Moreover, the human cognitive system is very sensitive to visible and concrete changes and finds it very difficult to react if the change is not easily perceived: this is a classic attentional phenomenon. This is precisely one of the problems of addictive behaviours that gradually degrade health, but so slowly and progressively that it is cognitively very difficult to react optimally. The parable of the frog that is boiled so slowly that it cannot perceive the temperature change necessary for the flight reaction has an undeniable psychological reality. Yet most environmental degradations, even if they are brutal in geological time, seem very slow or imperceptible to us cognitively. If the comparison of glacial surfaces photographed 30 years ago and today is so popular, it is because it allows us to better grasp what our cognitive system is incapable of grasping otherwise: the low visibility of the environmental problem we are facing is unlikely to encourage us to adopt the behaviour necessary for its resolution.

As for the second condition—the certainty that a change in behaviour can have an effect on the preservation of the common good—the situation is also much more complex than that suggested by Hardin in his thought experiment. In the case of Hardin's parable, we can imagine that the breeders form a small community of people. If their number is small enough, an individual decision can have a palpable effect on the common good. The individual impact is generally one tenth if the community includes 10 breeders. The problem is that we are nearly 8 billion humans on the planet. Not only are we too many to be able to exchange directly, but above all, we intuitively perceive that even a significant change in our individual behaviour will have little impact on the fate of the common good. This is the problem of the perceived incommensurability between individual action and collective action. If, on an individual level, I decide to reduce my consumption of fossil energy to protect the common good that is our atmosphere, my impact is 1 out of 8 billion: epsilon. If others do not adopt the same virtuous behaviour my decision will have no effect whatsoever except that of having reduced my own comfort of life. Now, what reason could lead me to suppose that the 8 billion humans, minus my modest self, will make the same decision as me and why would I sacrifice myself if I sense that others will not adopt the virtuous behaviour to which I have consented? This type of reasoning, which presents a certain rationality at the individual level but which is nonetheless absurd at the collective level, is a classic excuse that we all give ourselves to continue consuming as usual and thus actively participate in the destruction of the common good. Adopting virtuous ecological behaviour requires a sacrifice that one is all the more inclined to accept if one has some guarantees that others will do the same. It is trust in others that is at stake here. Now, all research shows that the level of trust that we place in others depends on their physical or relational proximity. It is very difficult for a human to place any trust in strictly unknown individuals living thousands of kilometres away and often perceived as potential competitors.

There is another major obstacle to altruism. We know that altruism is more likely to manifest itself when it is established between genetically close individuals: this is known as kinship altruism. It is in a way biologically constrained.[11] It is exercised gradually, weakening as the perceived distance between oneself and other individuals increases, this distance being largely determined by genetic proximity.[12] For example, parents are ready for total altruism to save one of their children in danger. This can lead them to the extreme sacrifice: giving their life to save that of the other. When genetic proximity decreases, altruism also decreases. If one has to choose between saving one's child and saving another child, the decision will be quickly made. If one has to choose between saving a child one knows and an unknown child, there will be little hesitation. Our ability to behave altruistically, as well as our ability to put our personal interest and our selfishness on hold, is directly related to the genetic proximity between oneself and the other. It should be remembered that until the Neolithic, humans lived in small groups generally not exceeding a few hundred individuals because neither villages nor even cities existed. Altruism was powerfully exercised within the group and it was exercised with all the more force because these individuals were mostly related. It is obvious that this form of altruism has little chance of being exercised in a community of nearly 8 billion humans genetically distant from each other.

The form of altruism that we have just discussed is undoubtedly the most archaic, the one that is constrained by the genetic proximity between individuals and which is strongly manifested in most animals. Fortunately, humans, and other cognitively evolved animals, have developed a form of altruism infinitely more complex, free from the requirement of genetic proximity.[13] This is reciprocal altruism: a form of altruism that can be exercised between people having no kinship ties as observed in the case of friendship for example. But again, reciprocal altruism does not set up under any conditions. We must never forget that altruism was first intended to be exercised from close to close simply because the human psyche evolved in a social universe systematically involving people knowing or rubbing shoulders with each other a little. It is considered that our hypertrophied brain is not capable of memorising the identities of more than 150 individuals (the average size of human groups in the Paleolithic) and therefore to establish social relationships based on trust and reciprocity if the group size becomes larger.[14] The current situation is problematic from this point of view: a humanity of nearly 8 billion humans, all strangers to each other and living in communities of gigantic size. The irony of Desproges' phrase, "there are more and more foreigners in the world", stigmatises the absurdity of racism and xenophobia, but nevertheless echoes the very real feeling of living in a community of billions of strangers. This is a fundamental difficulty because our altruism can hardly be expressed with strangers. The human population of the planet remains an abstraction that is not favourable to the triggering of altruistic behaviours, simply because direct interaction or communication between individuals is

[11] Hamilton (1964).

[12] Gavrilets and Fortunato (2014).

[13] Bowles (2006); Boehm (2012).

[14] Dunbar and Shultz (2007).

no longer possible. Not only is altruism less likely when it involves foreign people, but the very trust given to the other decreases with the level of perceived strangeness. This constitutes an additional reason not to adopt the altruistic and virtuous behaviour that our current situation would require. We have thus created a huge, impersonal and socially very divided, hardly conducive to the altruism that men are capable of in more traditional cultural environments.

This limit to altruism has been experimentally confirmed by psychologists and economists. Altruism, the trust given to others and the ability to coordinate by agreeing to put one's personal interest second, has been studied through game theory, a field of mathematics interested in strategic interactions between players with interdependent and sometimes competing objectives. The most popular example of these games is probably the prisoner's dilemma. In this game, the following offer is made to two suspects who are questioned separately.

> If you denounce your accomplice and he does not denounce you, you will be released and the other will get 10 years in prison. If you denounce your accomplice and he does the same, it will be 5 years in prison for each. If no one denounces the other, it will be 6 months for each.

In this game, if each follows a "rationally" selfish reasoning favouring his personal situation without taking into account that of the other, the probability is high that each will denounce his accomplice, the two suspects then getting 5 years in prison. By adopting this selfish behaviour, the result for each individual will be less favourable than the one that would have consisted in denouncing no one: 6 months in prison. Obviously, this other strategy assumes having great trust in one's partner because, in case of selfish behaviour by the other, the risk is to get 10 years in prison. Countless empirical research on the prisoner's dilemma and other related games, some of which precisely simulated the situation of the tragedy of the commons, have shown that individuals are much more cooperative than would be suggested by the classical theory in economics, according to which we are strictly governed by our personal interest. That said, the probability of cooperating or adopting altruistic behaviour in the context of a simulation of the tragedy of the commons is systematically reduced when (1) individuals do not know each other,[15] (2) it is impossible to "punish" those who would favour excessively selfish behaviour,[16] (3) individuals are too numerous,[17] (4) the decision is anonymous, (5) individuals cannot effectively exchange. So it is undeniable that altruism, as well as our rejection of inequity, are integral parts of the human behavioural repertoire, it is just as clear that having to interact with billions of humans who are perfectly anonymous to us is a major obstacle to the expression of altruistic behaviours. Nothing is

[15] Batson and Moran (1999).

[16] Fehr and Gachter (2002).

[17] Hannelore De Silva et al. (2010).

more difficult than escaping from one's natural selfishness when one's own interest is balanced against the interest of foreign individuals. It is far from obvious that any form of altruism can manifest itself between strangers.

7.4 The Challenge of Parochial Altruism

Altruism is an undeniably positive attitude and widely supported in all human cultures. However, altruism often has a very dark counterpart, capable of being at the origin of some of the most horrifying human behaviours. This includes a form of hostility, rejection, or even extreme violence, towards individuals identified as external to one's group of belonging.[18] Indeed, solidarity with members of one's own group often goes hand in hand with great hostility towards members outside the group, which we have called above parochial altruism.

In other words, like non-human primates,[19] we can be altruistic and put aside our personal interest when we interact with members of our small group. Similarly, we more spontaneously trust individuals from the same group as ourselves.[20] The tragedy of the commons is therefore all the more surmountable as it involves individuals who know each other and, above all, who perceive themselves as members of the same community. The great fashion for community life stemming from the hippie movement of the 60 s is undoubtedly the expression of a search for the optimal conditions for the development of altruism and cooperation. In humans, parochial altruism has taken on a unique dimension that is not found in any other animal species since we have managed to create virtual entities of individuals considered very close even when they are in reality perfect strangers to each other. This has notably allowed the emergence of a group consciousness that can include up to millions of individuals. Just as citizens of the same state, believers of the same religion, followers of the same sect, supporters of the same political ideology… During the great wars of the twentieth century, millions of individuals sacrificed themselves to "save" their homeland and therefore individuals they did not know and of whose existence they were totally ignorant. The devastating religious wars of past centuries are another disastrous manifestation of this. This is a particular case of cultural evolution that has led to the extension of altruistic and cooperative behaviour, initially biologically constrained to only be exercised with known members of a small community. Unfortunately, this is also what made possible large-scale national conflicts, capable of generating violence that has no equivalent in the animal world. Ordinary racism or xenophobia are only archaic behaviours resulting from a biological and cultural coevolution that led to large-scale parochial altruism. Today, nationalist leaders all appeal to national preference by playing on the strength of a parochial

---

[18] Sober and Wilson (1998).

[19] Wilson et al. (2012).

[20] Meloni (2014).

altruism that is expressed with all the more force as the feeling of insecurity and vulnerability increases. It is not difficult to understand that extended version parochial altruism is a real obstacle to the rational and efficient management of the common good because the hostility towards foreign groups, inherent in this form of altruism, makes large-scale coordination particularly difficult.[21] Unlike Hardin's parable, states are much more aggressive than breeders because parochial altruism is powerfully exercised within them. In the parable, the aim of a breeder was simply to enrich himself and not to enrich himself more than his neighbour, or even at the expense of his neighbour. In the current geopolitical context, there is much more than that. There is an economic war for leadership, a war that is manifesting itself today on multiple fronts. More generally, states are in competition with each other. As long as competition between foreign nations predominates over the altruism necessary for the management of common goods, as long as short-term economic interest remains their primary objective, the probability that the common good will be preserved remains very low. Trump's unilateralism or the rejection by European populist parties of supranational political entities are only contemporary forms of an archaic parochial altruism that constitutes a major obstacle to a global consideration of the tragedy of the commons. Need we remind that all attempts to reduce CO_2 in the atmosphere have systematically failed because some states believed that such measures would be detrimental to their economy, their competitiveness and therefore their hierarchical position on the geopolitical chessboard? As the Commission on Environment and Development acknowledged in 1987: "The Earth is unique but the world is not". A final problematic consequence of the competition between states is the absolutely absurd level of inequality to which it has led. The relationship between international rivalry and the level of inequality is not contingent because humans, like other evolved living species, develop organisations that are all the more unequal and hierarchical as intergroup rivalry is strong. The extreme inequality that characterises our contemporary societies is so familiar to us that it ends up appearing as a hardly contestable norm. But can we reasonably accommodate such an extreme level of inequality? It is now a real psychological obstacle to reciprocal altruism because a prerequisite for its triggering is the preservation of a level of equality or, at the very least, minimal equity. Applied to the political field, without a minimum of equity, it is impossible for the less wealthy individuals or states to adopt altruistic behaviours favourable to the planet, but unfavourable to their own situation.[22] In international negotiations on CO_2 emissions, the analysis of blockages reveals that the reasons for failure are almost exclusively related to the extreme level of inequality that radically divides states, particularly those in the South and those in the North.

[21] Ellemers (2012).

[22] Gsottbauer and van den Bergh (2010).

7.5 The Strength of the System

In these analyses, Hardin often referred to the "constraining system". At the heart of his analysis, therefore, was this powerful idea that humans have created a society that surpasses them and ends up imposing rules that may not be favourable to either individuals or the environment. To fully grasp its importance, it is useful to rely on a concept from cognitive science: emergence. We speak of emergence when the very numerous and complex interactions between the elements of a complex system give autonomy to the latter, conferring a constraining power on the elements that constitute it. When the level of complexity of the emerging system is sufficiently high, the system is, according to the established formula, more than the sum of the parts constituting it. It then manifests new, emerging qualities or properties that go far beyond those that characterise the parts themselves. Theories of emergence are particularly useful for accounting for what the human mind is.[23] Unless one adopts an outdated dualistic conception according to which it is our immaterial mind or soul that would be responsible for the characteristics from our psyche, it is indeed the 100 billion neurons of our brain that are the cause of this emerging reality that we call the human mind. Quite obviously, destroying all or part of the neurons of a brain leads to massive dysfunctions or to psychic death. Our most extraordinary psychic states, consciousness for example, result from the simultaneous activation of several billion neurons. Yet, a neuron is only a small biological structure devoid of consciousness, intelligence, desire, morality, empathy: all important characteristics of the human mind. It is therefore in this sense that we speak of emergence because the human psyche that results from the activity of these billions of neurons has properties that the neurons do not have individually. Theories of emergence have also been very useful for understanding the functioning of social insect societies such as ants, termites or bees. These societies result from the activity of millions of insects who, unknowingly and without having chosen, work towards the realisation of objectives that are those of the emerging system that is the colony. The ant hill is endowed with a collective intelligence that is much more than the sum of the intelligence of each ant.

If we accept the analogy, the system that we have created—our society—which results from the activity of more than nearly 8 billion humans would have its own autonomy, its own objectives without there necessarily being a thinking head or obscure leaders who would decide its trajectory. Yet, this latter is nothing other than the emergent product of the behaviour of billions of humans who lose a little more of their autonomy when the system itself gains it and ends up imposing on us what many reject individually. I am, like many, a perfect example. I have been for half a century very committed to ecology, I am part of those who contest the excesses of economic liberalism and yet, whatever the ideology that I defend, I fully participate in the system that I denounce. I can't even identify how I could extricate myself from it because the only solutions I foresee would be so radical and so

[23] Ripoll (2020).

psychologically and personally costly that I prefer to finally accommodate a situation that is rather favourable to me. Without being rich, I live well, I do not suffer directly from the degradations that we inflict on the planet and if it were not for the knowledge that I have in ecology and the moral values to which I adhere, I could very well continue to live my life ignoring the tragedy of the commons.

But what is the nature of the system that we have made emerge and why does it alienate us to this extent? This system was put in place as soon as the Neolithic revolution because already the technological and especially cultural changes that it induced have not, far from it, produced the surplus of happiness or individual well-being to which we aspire. The Neolithic society quickly took over the Paleolithic society not by virtue of the surplus of well-being that it individually bestowed on men, but by virtue of its efficiency and collective power: misery of the individual and power of the group as I previously titled it. The birth of an elite, a professionalised army, bureaucracy, religious authority, hierarchy, trade, inequality, poverty, expansionism… have contributed to the power of this new society not to the well-being of its population. That men, even the majority of them, paid a heavy price and saw their individual situation deteriorate had little weight relative to the power newly acquired by a system that they inadvertently brought about and of which they have become both dependent and submissive elements. Each hunter-gatherer who integrated this new culture, willingly or by force and whatever the reasons, was inevitably constrained to a new form of relationship to the world and to others. In a few generations, their culture was destroyed and, with it, their ability to live autonomously in their natural environment. This is the so-called tragedy of small steps: small changes and decisions that seem harmless taken separately propel individuals into a society of which they quickly become dependent, forced to adopt its way of life, without it being possible to return to the previous way of life. The same phenomenon is regularly observed again when, for example, Amazonian Indians are suddenly put in contact with loggers. In a few years, the introduction of money and a few consumer goods among the most useless (soda, cigarettes, trinkets of all kinds) are enough to destroy their culture and their autonomy. Having no choice, they must then adopt the new rules of the game, those that impose playing the game of competition with the aim of getting rich and achieving the social status that ensures them the best chances of survival. The system has taken over the individual who is then forced to adopt the implicit rules imposed by the society he has integrated unless he takes the risk of a marginalisation that very few humans are capable of accepting given our extreme sociability.

The more society becomes complex, the more powerful it becomes, the more it grows, the more the system becomes constraining and the more individuals become dependent on it, obliged to adopt a relationship with others and to the environment, problematic and painful for a large number of them. From this point of view, our society is the very example of an emerging autonomous system that imposes its objectives to which it becomes very difficult to deviate. The principles at the heart of the system that we have created are relatively simple. States must grow economically to ensure the material well-being of their citizens and have the military and technological power they need to protect themselves from competing states. Each individual naturally contributes to this growth through his desire to enrich himself and achieve the highest possible social status. That individuals proceed in a selfish manner to achieve their goal is economically rational because it is the selfishness of each and the work products to further enrich themselves individually, which contribute to the wealth of nations. Long before Smith, Mandeville's epic poem (1705) "The Fable of the Bees" had praised the selfish approach and the exclusive pursuit of personal interest in a society aiming for opulence and growth: "Be as greedy, selfish, spendthrift for your own pleasure as you can be, for thus you will do the best you can for the prosperity of your nation and the happiness of your fellow citizens". Mandeville was probably right about the prosperity of the nation and probably wrong about the happiness of the citizens because the goal of the system is indeed the prosperity of the nation, not the happiness of individuals. He had already identified the central role of the hedonic component—"spendthrift for your pleasure"—which should not be rejected, but on the contrary stimulated. He had formulated in simple terms the foundation of economic liberalism and Hayek himself considers him a precursor of the liberal theories of the twentieth century. Mandeville even thought that values still highly stigmatised by the Christianity of his time (luxury, vice, greed, the pursuit of immediate pleasure, the value of desire as a driver of action…) could magically transform into virtues and lead to collective prosperity through the simple mechanism of the market. Everything is said in this extract from the fable of the bees because if economics is a complex and sophisticated science, the economic behaviour of individuals in our society is governed by very simple principles well identified by Mandeville. A few days ago, an economic columnist from France Inter, a convinced neoliberal analyst, almost word for word repeated Mandeville's rhetoric, urging listeners who had saved too much because of the coronavirus health crisis to go and spend their money by returning to the shops. The implicit message he was transmitting was no longer to consume for pleasure and even less to satisfy some real needs, but to consume simply to restart the economic machine: the priority was no longer given to the individual, but to the system, the role of the individual being simply to support it by going to consume the money saved during the crisis. The injunction to consume was therefore justified by the need to satisfy the demands of a powerful system that can only survive if its growth remains fuelled by consumer behaviour. Economic reason is therefore primary and individuals must adopt behaviour compatible with economic reason whatever the consequences for themselves and for the environment: going to spend money in shops simply to keep the economy alive is an absurd injunction of which this otherwise intelligent and competent economist is no longer aware, being himself convinced that economic growth is the obligatory passage of any human society: alienation affects all men, including the most cultured and intelligent among them.

This prevalence of the system over the individual is perfectly illustrated by the organisation of social insect colonies.[24] An ant no longer has its own objective because its sole reason for being is to contribute to the overall health of the colony,

[24] Gowdy and Krall (2016); Rowland (2009).

its growth and expansion. Evolution has managed to do what no neoliberal thinker would have dared to hope or what no manager has managed to achieve with his employees because the activity of an ant is dictated by its genetic constitution. The very possibility of contesting the role assigned to it is totally evacuated. If it strays from the colony, it will die almost immediately because it is totally dependent on it: its alienation is total. It is nothing more than an element of a system, its autonomy as an individual being reduced to nothing due to the extreme specialisation inherent in the organisation of the colony. The individuals who defend the colony cannot survive without the individuals who feed and vice versa, just as the reproductive individuals cannot survive without those who feed or defend. The colony is an emergent system that follows its own objectives that the individuals themselves ignore. The system is perfectly constraining since it sanctions by death any individual who could not actively participate in its development.

Human society is "almost" as accomplished as that of social insect colonies. "Almost" because the hierarchical organisation into castes and functional specialisation is less radical in our society. Humans remain beings of consciousness, capable of contesting and questioning the system in which they are inserted. Nevertheless, as with social insect colonies, the organisation of society follows an ultimate goal: growth, the latter occurring all the more efficiently as each individual or group of individuals applies themselves to the pursuit of personal profit. If orthodox economists, whether liberal or not, have such contempt for the thesis of degrowth, it is because such a thesis simply consists in breaking with the ultimate goal of the system. However, such a questioning would generate such upheavals, with political and psychological implications so enormous, that many prefer the status quo, even if it is absolutely catastrophic in the medium term, rather than breaking with the system and its religion: growth. One does not break with one's alienation and with the objective that the system assigns to each of us just like that.

One of the most remarkable elements of the system, the one that is its most perfect expression, is the company, especially the large multinationals of today. Over the past two centuries, they have acquired a power and complexity that give them remarkable autonomy. They have become self-regulated systems, kinds of Frankensteins built by men, but ending up enslaving them to their sole and unique objective: profit. Their economic weight is such that states, themselves obsessed with national growth, are forced to pledge allegiance to them and often to yield to their demands. Their lobbies are so effective that they manage to bend states and their jurisdictions in order to eliminate any form of regulation that could hinder the maximisation of their profit. Whether this deregulation is detrimental to people is of little importance because their goal is not to make people happy, but simply to increase the level of collective wealth. The company has neither ethics, nor compassion, nor a sense of social justice nor respect for the environment. The autonomy of large companies is symbolically completed when the American court of justice considered that they could be considered as natural persons, full-fledged persons, while they are in reality only artificial constructions. By considering them as such, they benefit from inalienable human rights that are not subject to government regulation. They thus legally escape any attempt by the state to control them and benefit like any real human from the main amendments of the American constitution. They thus acquire the freedom to organise the lives of their employees in order to maximise their profits.

The company becomes an autonomous being, monstrously powerful and efficient in terms of wealth and profit production and we are all dependent on it. Its employees, whatever their role in the company, are naturally dependent on it to earn their salary without which they would be rejected by society. States, therefore the political leaders, depend on it to ensure economic growth which they need to perform their sovereign functions and to maintain the standard of living of citizens, probably also to maximise their chances of being re-elected. Shareholders, small or large, depend on it because a significant part of their income, especially at retirement in North America, results from the dividends that the company pays them. It is probably through this last point that the company can be truly identified as an autonomous emerging being on which shareholders depend. The company no longer belongs to anyone. It is now a collective being that belongs to millions of shareholders who will find it all the more difficult to oppose or challenge it because their personal financial survival depends on it. The separation between ownership, now distributed over thousands of shareholders, and the functions of direction and management, carried out by the board of directors, contribute to making it an abstract structure that escapes any form of control. Its intelligence is distributed over hundreds of thousands of employees who, like social insects, work with a single aim: to increase profit. Efficiency and rationalisation of work require that employees, even at high hierarchical levels, are highly specialised, their work being confined to repetitive and uncreative tasks. Taylorism has allowed huge productivity gains at the cost of a terrible alienation that is increasing today given the possibility of substituting robots for humans.

Because the company is evaluated almost exclusively on its profitability, the consideration of other parameters, human or environmental, becomes totally secondary except when the need for corporate communication requires displaying a surface environmental or social sensitivity. The requirement for profitability is necessary to satisfy shareholders as it is necessary for the survival of the company since generalised competition will sound the death knell for any economically underperforming company. Employees have integrated this so well that they are ready to earn less and work more to save the monster on which they depend. The monster is so powerful and demanding that it leads individuals to behave in a way that they themselves would condemn if they were not under its influence, thus generating painful states of cognitive dissonance. They must thus sacrifice their sensitivity, sacrifice what seems essential to them as a conscious human being, abandon the values they hold to adopt those of the corporate culture. This situation concerns even more managers than workers because they have the mission to impose on others what the company demands to maximise its profitability. Managers become a kind of extension of the monster by proceeding to a scientific organisation of work that is devastating on a human level, but economically efficient. This often results in a severe form of psychological dissociation, the same individual can be altruistic, concerned about others or the environment in the private sphere and adopt a behaviour perfectly opposed within his company. Turning a blind eye to human maltreatment, ignoring the ignoble exploitation of workers in developing countries, rationalising as best they can an unacceptable management style, putting up with disastrous ecological or health consequences, are just forms of denial without which the manager could not survive. The company thus gradually erodes the part of humanity that remains in each of us and also reduces the part of autonomy that still distinguishes us from social insects. It alters the meaning that we can give to our lives and which is so necessary for mental health.[25]

In a few decades, the financial weight of these large companies has continued to grow. Today the GAFAM alone represent more than 4000 billion in stock market valuation, a value higher than the GDP of Germany and much higher than that of France. Globally, 41,000 companies are listed on the stock exchange for a capitalisation of 84,000 billion dollars, equivalent to the global GDP.[26] These financial monsters have acquired a power all the greater as states, confronted with the systemic problem of debt, depend on them to support their economy. This situation is not contingent because, in a world of competition where the cardinal value is that of profit, states cannot compete with the company for several reasons. The company's purpose is to make a profit. It is designed for this, not states. The company escapes the democratic sanction of citizens, not states: we hardly see mass demonstrations against Facebook or Microsoft. Moreover, the company is not a clearly identified entity. It is indeed an emerging system of which we are all micro-parts, as employees, as shareholders and as consumers. To take only the example of Amazon, there are more than half a million employees, several billion Amazon customers and several hundred million people who directly or indirectly hold Amazon shares. Each individual now maintains a dependency relationship with the company. We can criticise them, we can be exasperated by the fact that they organise the march of the world, we can deplore that they are only driven by the maximisation of profit, but in reality, we are largely in solidarity with and dependent on them, as employees, as shareholders, as consumer customers. We are constituent elements of the monster we have brought into being and we are all the more so as at the individual level, almost every one of us tries to maximise our profits and accumulate more wealth.

The disastrous consequence of all this is that the objectives of the company are neither to increase human well-being, at most to increase their level of wealth, nor to preserve the environment. Therefore, if their power continues to increase, there is no chance of reaching a favourable outcome to the tragedy of the commons. Hardin was quite right from this point of view. If the emerging systems that are companies escape our control, if they escape any form of political regulation, if their autonomy is such that they manage to constrain even our individual behaviour, then the human anthill will continue its expansion blindly until the moment when, the earthly limits being crossed, we will be brutally confronted with environmental problems that will

[25] Bohler (2019).

[26] Who owns the world's listed companies and why should we care? [archive], OECD, 17 October 2019.

force us to put an end to a deadly system that is a few 10,000 years old. A system of formidable efficiency, which has increased the wealth of billions of humans, which has allowed the emergence of art, philosophy and science, but which at the same time has degraded the living conditions of other billions of humans, ravaged the terrestrial ecosystem and led to a level of inequality that is totally astonishing. The whole question is whether we will be able to find an alternative to this model before the degradation of the ecosystem is such that life on Earth becomes impossible.

References

Alberto, M., & Ravat, J. (2011). *Human morality and sciences*. Editions Matériologiques.

Batson, C. D., & Moran, T. (1999). Empathy-induced altruism in a prisoner's dilemma. *European Journal of Social Psychology, 7*, 909–924.

Boehm, C. (1997). Impact of the human egalitarian syndrome on Darwinian selection mechanics. *American Naturalist, 150*, 100–121.

Boehm, C. (2008). Purposive social selection and the evolution of human altruism. *Cross-Cultural Research, 42*, 319–352.

Boehm, C. (2012). *Moral origins: The evolution of virtue, altruism, and shame*. Basic Books.

Bohler, S. (2019). *Le bug humain*. Paris, Robert Laffont.

Bowles, S. (2006). Group competition, reproductive leveling, and the evolution of human altruism. *Science, 314*, 1569–1572.

de Waal, F. B. (2008). Putting the altruism back into altruism: The evolution of empathy. *Annual Review of Psychology, 59*, 279–300.

Denault, L. K., & McFarlane, D. A. (1995). Reciprocal altruism between male vampire bats, Desmodus rotundus. *Animal Behaviour, 49*, 855–856.

Diamond, J. (2005). *Collapse: How societies choose to fail or succeed*. Viking.

Downey, S. S., Randall Haas, A., & Shennan, S. J. (2016). European Neolithic societies showed early warning signals of population collapse. *Proceedings of the National Academy of Sciences, 113*(35), 9751.

Dunbar, R., & Shultz, S. (2007). Understanding primate brain evolution. *Philosophical Transactions of the Royal Society B, 362*, 649–658.

Ellemers, N. (2012). The group self. *Science, 336*, 876–879.

Fehr, E., & Gachter, S. (2002). Altruistic punishment in humans. *Nature, 415*, 137–140.

Gavrilets, S., & Fortunato, L. A. (2014). Solution to the collective action problem in between-group conflict with within-group inequality. *Nature Communications, 5*, 3526. https://doi.org/10.1038/ncomms4526

Gowdy, J., & Krall, L. (2016). The economic origins of ultrasociality. *The Behavioural and Brain Sciences, 39*, 1–60.

Gsottbauer, E., & van den Bergh, J. C. J. M. (2010). Environmental policy theory given bounded rationality and other-regarding preferences. *Environmental & Resource Economics, 49*(2), 263–304.

Hamilton, W. D. (1964). The genetical evolution of social behaviour. I & II. *Journal of Theoretical Biology, 7*(1), 1–52.

Hannelore De Silva, H., Hauert, C., Traulsen, A., & Sigmund, K. (2010). Freedom, enforcement, and the social dilemma of strong altruism. *Journal of Evolutionary Economics, 20*, 203–217.

Hardin, G. (1968). The tragedy of the commons. *Science, 162*(3859), 1243–1248.

Hutchings, J. A., & Reynolds, J. D. (2004). Marine fish population collapses: Consequences for recovery and extinction. *Bioscience, 54*, 297–309.

Meloni, M. (2014). How biology became social, and what it means for social theory. *The Sociological Review, 62*, 593–614.
Ostrom, E. (1990). *Governing the commons: The evolution of institutions for collective action (Political economy of institutions and decisions)*. Cambridge University Press.
Ripoll, T. (2020). *Pourquoi croit-on ? Psychologie des croyances*. Auxerre, Sciences Humaines.
Rowland, W. (2009). Reflections on metaphor and identity in the cyber-corporation. *Journal of Business Ethics, 90*, 15–28.
Sober, E., & Wilson, D. S. (1998). *Unto others: The evolution and psychology of unselfish behaviour*. Harvard University Press.
Van Veelen, M., & Hopfensitz, A. (2007). In love and war: Altruism, norm formation, and two different types of group selection. *Journal of Theoretical Biology, 249*, 667–680.
Wilson, M. L., Kahlenberg, S. M., Wells, M., & Wrangham, R. W. (2012). Ecological and social factors affect the occurrence and outcomes of intergroup encounters in chimpanzees. *Animal Behaviour, 83*, 277–291.

Chapter 8
But What Kind of Ecology Are We Talking About?

Abstract Ecological thought can take many forms and be claimed by very different political currents, ranging from the far left to the least democratic and most totalitarian far right. Similarly, philosophers have developed radically different conceptions of ecology. In this chapter, I try to produce an analysis grid of these different currents. I am particularly interested in the magical and esoteric conceptions of ecology and contemporary transhumanism and its naive belief that scientific progress will allow us to solve environmental problems without the need to question the economic ideology of the necessity and possibility of infinite growth.

Ecology, as a political approach, has become omnipresent, both in bar discussions and in the strategic and electoral speeches of politicians or major business leaders. If it now constitutes a fundamental element of our ideologies, it manifests itself in extremely heterogeneous forms, both taken up by left-wing, far-left, right-wing and far-right parties. This polymorphism has two major disadvantages: it suggests an opportunistic exploitation of ecology that can end up tiring or even irritating, and it leads to a form of dilution and fragmentation of ecological thought that can only weaken it at a time when it should be imposed. The objective of this chapter is to expose these different positions and to affirm the one that seems to me to be the most capable of evolving our societies so that they can survive the environmental crisis. The great difficulty in this exercise is to avoid caricatures and amalgams as political ecology is teeming and complex and as it is crossed by very different political conceptions.

T. Ripoll, *Human Psychology and the Destruction of the Planet*, Anthropocene – Humanities and Social Sciences, https://doi.org/10.1007/978-3-032-02080-2_8

8.1 Some Philosophical Points of Reference

Many authors consider[1] that the relationship that men have established with nature, the very one that has led to its destructive exploitation, has its origin in what is at the foundation of Abrahamic religions: anthropocentrism and a representation of time that is at the origin of the concept of progress. Religion would therefore have constituted a conceptual soil favourable to the predation that we have exercised on our natural environment.

Anthropocentrism, at least the superior value that we give to human compared to the rest of the environment, results from a radical form of dualism, the metaphysical base of all monotheisms. Men would be endowed with a spirit and a body while other living beings, a fortiori material objects, would be strictly material beings. In this sense, the three great monotheisms broke with animism that prevailed in the Palaeolithic and early Neolithic periods, since among animists, objects as well as living beings are endowed with a soul. Spirituality, in the broad sense of the word, is evacuated from nature. The latter has lost the sacredness that men accorded to it before the great religions imposed themselves. Nature is subject to strictly material determinisms that science can explore and master. Spirituality is solely the prerogative of God, and of course of Man, God having created us in his image. This duality, first between matter and spirit and then between human and nature, is an ordered duality. The spirit is superior to matter to the point that matter, especially the human body and the desires associated with it, are perceived as a dangerous remnant of our animality by most religions. Consequently, human and nature are certainly not on an equal footing either. Nature is a part of God's creation whose essential function is to meet human needs, a classic expression of unfounded teleology. Some well-known extracts from the Bible are very revealing on this subject:

> God created human in His own image, in the image of God He created him; male and female He created them. God blessed them, and God said to them, 'Be fruitful and multiply, fill the earth and subdue it; and have dominion over the fish of the sea, over the birds of the air, and over every living thing that moves on the earth'.[2]

Man is definitively separated from nature. He is its crowning glory. He is made in the image of God and must therefore be distinguished from other living beings. A new ontological chasm thus creates an infinite distance between nature and us. In a certain way, in a mystical and explicit form, these religions have only affirmed what guides the life of every living being: to reproduce in order to ensure the survival and expansion of the species, a form of genetically constrained parochial altruism which consists in prioritising one's species above all others. Man's exploitation of nature and its inevitable destruction could not pose a problem since God created nature for us to dispose of and for us to reproduce infinitely. This extract from Genesis (9, 1–3) expresses it not without some brutality:

[1] White (1967); Metzner (1994); Zarzan (2005).

[2] Genesis 1:26–28.

> Be fruitful, multiply and fill the earth. Be the fear and dread of all the animals of the earth and all the birds of the sky, as well as everything that teems on the earth and all the fish of the sea: they are delivered into your hands.

Much more recently, in a text from the Vatican in 1965, similar remarks are made:

> For believers, one thing is certain: considered in itself, human activity, individual and collective, this gigantic effort by which men, over the centuries, strive to improve their living conditions, corresponds to God's design. Man, created in the image of God, has indeed received the mission to subdue the earth and all that it contains, to govern the cosmos in holiness and justice and, in recognising God as the Creator of all things, to refer his being as well as the universe: so that, while being subject to man, the very name of God may be glorified throughout the earth.[3]

This anthropocentrism and the little consideration given to nature have naturally been the target of many ecological thinkers who have identified religion as a major cause of the current environmental catastrophe. I would be a little more circumspect here. Certainly, we cannot deny the obvious convergence between the position of religion regarding the Man/nature relationship and the little respect that we have given to nature for over twentieth centuries. From there to consider a strong causality, there is a step that I would not take. It is quite possible that religion, at the time of the biblical writings, did nothing more than express the representation that men had of their relationship to nature. Religion is a cultural production of men who have conceptualised a relationship to the world that has prevailed because historically it could not be otherwise. The fact that the Vatican has significantly evolved its doctrine in recent years shows that religion tends to conform to the ideological evolution of the moment. It is undeniable that many believers have managed to extricate themselves from an extreme form of anthropocentrism. Pope Francois himself has converted to ecology as evidenced by his encyclical, even though he concedes in private that "before I didn't understand anything", an admission full of humility that illustrates the recentness of the ecological revolution that has taken place at the Vatican. Many moderate believers are now sensitive to ecology while maintaining a watered-down form of anthropocentrism according to which we must take care of nature, not because nature has an intrinsic value, but because its preservation is necessary for the well-being, development and survival of man. To quote the words of Bertrand Sajaloli and Etienne Gresillon,[4] the Vatican has moved from a conception of human as master of nature to a conception of human as guardian. This is still a utilitarian conception of nature, the one that was clearly expressed in Genesis, but which has undergone a remarkable inflection. In reality, the Catholic Church, at least a part of it, has undergone a real revolution, to the point that more and more conservative voices are rising against the radical positions of Pope Francois: a Pope decidedly humanist, challenging consumerism and liberalism and

[3] Constitution *Gaudium et spes*, verse 34, paragraph 1. Text in French on the official Vatican website.

[4] Sajaloli and Grésillon (2016).

advocating a form of joyful austerity that strangely echoes the positions of the proponents of degrowth.

The second fundamental characteristic of the great monotheisms lies in their conception of time, all imbued with teleology: a messianic conception admitting that human must evolve towards a better future. Time is no longer cyclical, as it was in older conceptions; time is linear and progress is constitutive, even imperative. From the seventeenth century, the notion of progress secularises due to the development of science and technology. The spiritual development to which religion aspired is replaced by a scientific development which it is thought will increase the happiness and well-being of all humans. Earthly happiness replaces heavenly salvation and becomes the main objective that all humanity should follow according to a clearly affirmed universalist vision. It is in the nineteenth century that the ideology of progress reaches its peak, with science being perceived as the most noble human activity, the most representative of its genius, the one that will best distinguishes it both from other living beings and from "archaic" human peoples who have not yet crossed the threshold of scientific rationality. Progress is first perceived as scientific and technical. But it is also, in continuity with the philosophers of the Enlightenment, conceived as political and intellectual. Political because progress only makes sense if all humans benefit from it and if it develops within democracies capable of taking advantage of it, with the aim of establishing an open, free and socially just society. Intellectual in the sense that the development of science must contribute to the eviction of all obscurantism, to the development of rational and critical thinking.[5] A part of humanity, the West, then experiences a feeling of boundless power and accomplishment, that of having emerged from barbarism and obscurantism, that of being the holder of a model of civilisation that should impose itself on the whole planet. This wonderful and naive optimism will unfortunately begin to crack at the end of the nineteenth century and especially during the twentieth century for good and bad reasons. It is clear that the dramatic wars of the twentieth century, the realisation, with the atomic bomb, that scientific progress could lead to the destruction of the planet and that science is sometimes at the service of states in their will to power and their expansionism, have largely contributed to the distrust that many express today towards science and progress as a whole. The realisation, in the second half of the twentieth century, that scientific progress associated with productivism, industrialism and liberalism, is a direct and indisputable cause of the ecological disaster, has finished destroying the perhaps naive belief in a linear progression of humanity towards an ideal future. It is very likely that we are the first humans to fear that the future may be much darker than the present.

In a certain way, the main current political ideologies result from the treatment they operate on the two historical foundations of the great religions that we have just recalled: anthropocentrism and the belief in the existence of constant progress. It is in the rejection or acceptance of a conception of progress all oriented towards the sole well-being of humans that will be drawn very contrasting philosophical and political conceptions, sometimes completely antinomic. Within them, ecology will take very different forms that we must now expose.

[5] Sanchez (2007).

8.2 Ecology in the Far-Right and Traditionalist Liberal Right

This is a minority current in France, clearly represented by a part of the traditionalist far right and the most conservative liberal right. In relation to the history we have just made concerning the impact that religion has had on our conception of nature, these politicians or intellectuals do not question either anthropocentrism or the belief in the virtue of progress. Their attachment to the maintenance of the current liberal economy, by ideology, by defence of their own interests or national interests, is so powerful that many more or less explicitly deny the dramatic dimension of the planet's environmental situation. It is among these politicians or these "intellectuals", whose most remarkable and caricatural representatives are mainly found in the USA, that there are the largest number of climate skeptics. Essentially, these are conservatives who perceive ecology as intimately linked to socialism and therefore likely to generate a form of revolution that would be detrimental to them. If such a movement remains relatively minority in Europe, it remains nonetheless quite powerful, especially in the USA, the country which remains the one that still sets the tone for the world's progress. This movement represents the zero level of ecology. With a few remarkable exceptions, their leaders are characterised by a very low level of scientific knowledge of current environmental problems. I do not think that this type of movement has much future, which is why it often operates in a masked way as it is every day more difficult to deny an empirical and scientific evidence that is increasingly striking. However, because of the financial or political power of its members, it is a brake on the rational consideration of the ecological problem.

8.3 The Progressive Neoliberal Right and Its Caricature: Libertarianism and Transhumanism

This movement is both close to the previous one and very distant. It is close in the sense that anthropocentrism is probably pushed here to its extreme, not because of the particular link we would have with God, but because of the sole "genius" of man, notably his ability to transcend his own limits and those of nature. It is also close because scientific and technological progress is at the heart of its vision of the future. However, it differs radically because it rises against any form of tradition and conservatism. Politically, states are perceived only as obstacles to individual freedom, which should be maximal. They no longer represent anything more than outdated political and technocratic obstacles that can only slow down the progress of a progress always developing in proportion to the freedom, granted to individuals or companies, to innovate and profit from their creativity.

Within this ideological nebula with multiple branches, ecology is not ignored, the importance given to science in this movement does not allow them to deny the reality of the ecological crisis. However, their proposals to solve this crisis are either purely corporate communication, or a disturbing utopia inflated with a *hubris* pushed to its peak. To take just one extreme example among many others, how can we take seriously the tears that Elon Musk sheds about the climate crisis when, at the same time, this scientist and businessman pollutes what remains of the most extraordinarily natural and beautiful on our planet, the sky, by putting these satellite caterpillars into orbit to spread broadband across the entire planet, or when he sends a Tesla car into space in a project where smoky scientific perspectives and crude corporate communication cleverly merge? Not to mention the Neuralink project, and its goal of implanting a connected chip in a human brain, both intended to treat neural pathologies, addictions, depressions, to endow the brain with superior cognitive power, to store memories in digitised form and, why not, to download them into another body or into a robot in order to endow us with a form of digital eternity. As Musk himself concedes in a burst of humility and caution that contrasts with his usual arrogance "the future is going to be weird" and we should not let "AI turn men into domestic cats of our computers". And then, to avoid the extinction of our species (what about our environment by the way?), Musk suggests migrating to Mars, leaving behind a polluted and unlivable planet. What a dramatic solution that reveals a very weak attachment to the beauty of terrestrial nature! As Musk also admits:

> It's going to be hard. There's a good chance of dying on the journey in the small capsule through space, and once you've landed successfully, you'll have to work non-stop to build a base (…). It's a very hostile environment.

Obvious and delirious excess that would prepare us for a dark horizon before even nurturing the slightest project capable of sparing us a one-way trip to a planet on which life can no longer naturally develop. Let's take a more serious look at the compatibility of this extreme form of belief in progress with the resolution of the ecological crisis.

What is absolutely fascinating about this story is its antiquity, as if humans had sensed well before the emergence of contemporary science and technology, the dangers of wanting to transgress the limits of our humanity and those of our environment: the danger of excess and the religion of progress associated with it. The myths of Prometheus, and more recently of Frankenstein, are perfect examples. In transhumanism, then post-humanism, the excess is twofold. First, because of technological development and the goal of infinite growth, human is heading towards such massive destruction of his terrestrial environment that the prospect of living there normally seems increasingly unlikely. Two paths then open up to him. Either human manages to curb the deteriorations he has made thanks to further technological and scientific progress—this is the sometimes delirious programme of cornucopian eco-engineering—or, more radically, it is a question of modifying human and enhancing him cognitively and biologically, so as to enable him to survive in an environment that has become hostile due to the degradations we have inflicted on it. In the event that neither one nor the other is possible, there remains the both unrealistic and terrifying fantasy of cutting ourselves off from our natural environment and going to live on another planet.

We are here at the opposite of the classical approach in ecology developed by Hans Jonas[6] in his principle of responsibility, that is to say the imperative necessity to be cautious in terms of technology and to adopt the precautionary principle with a view to not harming either the environment or future generations. On the contrary, for transhumanists, there is a moral right to use technology to increase our cognitive and biological capacities in order to become master of life, just as there is a moral right to infinitely modify our environment to adapt it to a human evolution for which we are now responsible. If human needs to be modified to survive in the new environment he has created, then everything must be done, including the possibility of creating an enhanced man, evolving him or selecting him appropriately. Why should we accept the natural laws of evolution when we have the means to create the world we want and the superman who will be its master? Presented in this way, transhumanism probably appears as a disastrous and worrying declination of the human will to substitute itself for the natural order, revealing a form of will to power that sometimes belongs to the most banal psychiatric pathology. But in reality, transhumanism, beyond these excesses, poses a real and very complex problem, more subtle than it seems and whose resolution can have great consequences for the future of political ecology.

8.4 The Problem Posed by Transhumanism and Its Watered-Down Form: Cornucopianism

Transhumanism is an extreme form of the cornucopian approach. Under infinitely more transgressive and dangerous appearances, it differs little in reality from cornucopian orthodoxy and its pseudo-rational belief that infinite growth is possible provided we have suitable technologies. The belief in so-called sustainable development, in green growth or the belief that the development of renewable energies will be a solution to the problems we encounter, presents a veneer of rationality, measure and moderation that should not be deceiving. Behind the facade, there is first of all the will to maintain unchanged a politico-economic system. Out of breath.[7] To some extent, the cornucopian approach is even more dangerous than transhumanism because, if the latter can rightly frighten us due to its obvious excess, the former is more insidious, wanting to persuade us that we can continue to live "normally" thanks to the upcoming progress in science and technology. To fully grasp the obvious absurdity of such a belief, the most convincing consists in placing oneself in the extreme case where we would have an energy renewable to infinity and perfectly non-polluting: the cornucopian grail supposed to allow us to solve all environmental problems. This is exactly the nuclear fusion project, the ITER project, currently pursued at the Cadarache site in the South of France. The obstacles usually

[6] Jonas (1979).

[7] Feygina (2013).

mentioned by opponents to this pharaonic project, swallowing up billions of euros, are of two types. On the one hand, they denounce the possible non-feasibility, at least the very hypothetical, of a project that consists of nothing more or less than trying to reproduce in a "controlled" way the nuclear fusion processes as they occur at the level of the sun. On the other hand, they denounce the environmental risks inherent in a technology still in its infancy whose mastery remains very uncertain in view of the absolutely phenomenal energy that will have to be produced and controlled. From my point of view, these criticisms, as reasonable as they may be, are not are not as fundamental as that. After all, many technologies once new and worrying are now perfectly mastered and bring us undeniable benefits. The problem is elsewhere. It is technical and psychological.

Technically, what would become of our planet if we had such energy? The problem of climate change would be perfectly solved since there would be no or almost no emission of CO_2. Obviously also, such energy would allow a less polluting recycling and therefore would globally reduce our ecological impact. But it is quickly forgetting the devastating and inevitable effect of the rebound effect, a psychological effect that leads men to systematically increase their consumption as soon as a technology allows to reduce the cost of energy and its environmental impact. In fact, any energy saving linked to any technological progress leads to an inevitable increase in consumption inevitably leading to the continued destruction of our environment. But of course also, such energy would boost our economy. It would generally raise our level of growth and our level of consumption. The global standard of living of humanity in would be significantly improved. To take just two examples, it is likely that, as the standard of living increases, the global population would increase and with it the need to produce more agricultural commodities. However, agriculture, apart from any energy aspect, contributes greatly to the destruction of the planet either through the chemistry it uses to fertilise the soil or limit pests, or through the destruction of natural spaces as is tragically the case currently in the Amazon or in Indonesia. The second example concerns the enormous amount of raw materials that we extract to produce everyday consumer goods directly contributing to the destruction of our environment. Energy is not everything. Its lack can be a limiting factor, but its abundance, even if it is clean and renewable, would not change much to the fact that any infinite growth will always be confronted with the obstacle of the finiteness of our planet. Another way to become aware of the illusion of green growth supported by the existence of clean and infinitely renewable energy is to imagine what our planet would be like if we had had this miracle energy since the Neolithic. Obviously, the planet would be in a worse state than it is today because this infinite energy would have contributed to boost our growth even more and would not have changed our ecological footprint in any way. The only thing that would have differed would have concerned our CO_2 emissions and therefore the issue of global warming which is in fact only one problem among many others. The current loss of biodiversity for example has been very little impacted by the fact that we use fossil energy: it would not have been much different if we had had the energy produced by nuclear fusion. Ideology is so blinding that it sometimes leads us to a form of magical thinking the most irrational. How can we still believe that our growth will not harm the planet as long as we grow green?[8]

8.5 A Credulity That Has Multiple Causes?

Why don't we want or can't we admit that technological innovation and human genius cannot succeed in overcoming the ecological crisis in an economic and political context of infinite growth?

The first of these causes is related to the economic competition that states wage, but also to the fact that citizens themselves, whatever their sensitivity to ecology, continue to evaluate their leaders on the basis of macroeconomic results—GDP for example and the growth rate—and their consequences in terms of unemployment and purchasing power.[9] The conjunction of these two factors constitutes a pressure to which no political leader can escape. Losing the economic war is synonymous with slowing down growth, therefore increasing unemployment and reducing degradation of citizens' standard of living. The overall pressure of the system is such that the margins for manoeuvre in terms of ecology remain small. The only measures that will be taken are those that are likely to contribute to growth or, at the very least, not to reduce it. The rhetoric recently used by Emmanuel Macron on the implementation of 5G is very symptomatic. In front of a group of entrepreneurs from the "French tech", the President clearly took a stand in favour of 5G*:*

"France is the country of Enlightenment, it's the country of innovation (...) We're going to twist the neck of all the false ideas." *or* "I hear many voices rising to explain to us that we should address the complexity of contemporary problems by going back to the oil lamp! I don't believe that the Amish model can solve the challenges of contemporary ecology."

Beyond the caricatural rhetoric of making a dubious analogy between the Amish movement and the demand from part of the left to apply the precautionary principle through a moratorium, the President, by invoking the rationalist and progressive conception of the Enlightenment, defends the implementation of 5G simply because he fears that any delay in this type of technology risks handicapping France in international competition. It is quite possible that he adheres to a form of cornucopianism, but even if this were not the case, the requirement for short-term economic success leaves him little choice but to delay the implementation of 5G. Economic warfare is really one of the most obvious obstacles to the adoption of real ecological measures. To use an analogy that will irritate the world of cycling and sport, if it's about winning the Tour de France, in the absence of effective medical control, there are no other solutions than to dope. Competition, little or unregulated, always leads

[8] Bjorn et al. (2018).

[9] Costanza et al. (2014, 2014); Drews (2017); Kallis (2011).

to dangerous behaviours. If a cyclist decides not to dope—not to equip himself with 5G—, he will lose the Tour de France—he will lose the economic war—and will have little chance of benefiting from a new contract—of being re-elected for a political leader. Under these circumstances, doping is inevitable as is the use of technologies which, even if they are potentially dangerous for the planet and for humans, can contribute to national economic growth.[10] As long as international competition remains the major determinant of our national behaviours, it is unlikely that ecology will be placed at the heart of national policies: this would be economically and socially suicidal. As for defending an extreme form of isolationism and protectionism that would protect us from the rigours of fierce international competition, this is no longer possible as our interdependence is now total. This results in a well-known defence mechanism in psychology consisting of rationalising an irrational or morally condemnable behaviour from pseudo-rational justifications. This is exactly what happens when our leaders come to persuade themselves that technological progress will save our planet and our economy.

The second cause lies in our conservatism. This may seem contradictory insofar as cornucopianism, and a fortiori its extreme form, transhumanism, precisely claim to work for a constantly renewed world and certainly not for a form of status quo or even worse regression. This is indeed the meaning of the criticism that President Macron addresses to those who are wary of 5G, a criticism echoed by so many others according to which we cannot meet the challenges of the contemporary world by "going back to the oil lamp". Despite appearances, it is indeed a conservatism that ignores itself in the sense that just as our societies cannot survive without growing, they cannot survive without evolving and without innovating: innovation is an integral part of our system. We must acknowledge the fact that there is a deep convergence between economic liberalism and scientific innovation for the simple reason that liberalism, because it places competition between individuals and states at the heart of its system, also conceives of innovation as the ultimate weapon without which one loses the economic war. Moreover, innovation is at the heart of our system because it constitutes a perfect incentive to consumption and therefore conditions the growth we cannot do without. In other words, to preserve the current system, for it to remain largely unchanged, we must continue to favour innovation over any other type of concern. This form of conservatism has a real psychological dimension and is linked to the so-called momentum effect bias well known to economists. This is an extremely powerful effect that leads traders, for example, to invest confidently in investments for the simple reason that they have been profitable in the past. There is no doubt that we operate in the same way with regard to the ecological crisis and its possible resolution through scientific innovation. Even when we rationally perceive the looming catastrophe and the incompetence of the system that leads us to it, we remain somewhat confident that technology will save us. The reasoning, insofar as we can talk about reasoning, is simplistic but powerful. Since

[10] It is highly likely that 5G does not pose a real danger, but this is secondary in light of this argument.

for 10,000 years, humans have always found technical solutions to the problems encountered, since we continue to live normally while we have been alerted for 50 years to the major dangers we are running, why wouldn't things continue this way? Even if one is not a climate skeptic and a convinced cornucopian, a part of ~~we~~ us probably continue to believe that humanity and its genius will find a technical solution to the environmental problem.

The third cause is not unrelated to the momentum effect. It has been particularly well studied by cognitive psychologists, notably Kahneman and Tversky. We are generally victims of a so-called optimism bias. This bias consists of generally overestimating our future success capabilities because we systematically believe that our goals are more achievable than they actually are: we are generally unrealistic about our predictions for the future. It also consists of overestimating ourselves. Thus we generally believe we are more likeable than others, more insightful, more competent than our peers, less dependent on cognitive biases and less sensitive to negative influences: this is the so-called self-illusion phenomenon. This bias is not only negative because without a minimum dose of optimism, we would probably not undertake and succeed in anything. This is well expressed by Mark Twain's amusing phrase: "They didn't know it was impossible, so they did it". But such a bias is also extremely dangerous as it distorts our appreciation of reality. It is particularly catastrophic among teenagers who, almost magically, believe they can engage in risky behaviours more than others without incurring real dangers. Typically, teenage smokers greatly underestimate the increased probability of cancer related to smoking even when explicitly informed of the risk. Such a bias is extremely common among traders who can no longer accurately assess their risk-taking. In a study conducted by Kanheman,[11] entrepreneurs were asked to estimate what percentage of businesses were able to survive 5 years after their creation and the chances of their own business surviving after the same period. Their estimates are respectively 60% and 80%. They perceive themselves, in accordance with the optimism bias, as more competent and efficient than others, because, in reality, in the USA, only 35% of small businesses survived 5 years after their creation. This powerful bias does not explain everything of course. But it helps to understand why, despite undeniable empirical data, despite the physical impossibility of growing indefinitely in a finite world, very intelligent individuals can be so stubbornly blind.

The fourth cause is also psychological. It is also the most powerful, the most complex and the least controllable because it is deeply rooted in our psyche. It results from the conjunction of our extraordinary cognitive abilities, our infinite quest for happiness and completeness—therefore the awareness of our limits—and the naive belief that changing our material living conditions will contribute to our happiness.

This is where the Neolithic cultural revolution played a significant role. It was the civilisational trigger that definitively placed humans on a trajectory that is difficult to control since it is constantly evolving towards a future that is always

[11] Kanheman (2011).

different and theoretically better than the present. Of course, over the million years preceding the Neolithic, major technical evolutions took place (mastery of fire, manufacture of the first stone and bone tools, transition from chipped stone to polished stone, first arrowheads, scrapers and first daggers…), but these were so diluted in time that it is likely that humans at the time were not even aware of witnessing a real innovation. The Neolithic instilled in the human psyche the certainty that the human condition was not fixed once and for all and that it could evolve in proportion to the ability to concretely intervene in the world, on its environment, on others: in short, to innovate.

Most researchers, anthropologists, neuroscientists and psychologists,[12] who work on human creativity agree that its explosion during the Neolithic resulted from the conjunction of a biological evolution of the brain and a major cultural revolution. A determining element of the cultural evolution of the Neolithic that revolutionised human life is the fact that we were led to live in villages and then in cities and networks of increasingly large cities. This is not insignificant because innovation is strictly linked to the frequency of our exchanges with others. The more we are integrated into large networks of humans, the more we interact with them, the more creativity increases. Not only does this contribute to emulation, but being eminently social beings, we are continually inspired by others to create. Solitary creation is a romantic myth without foundation. Cultural innovations are like viruses, they spread better and mutate faster as men form dense networks. From this point of view, our globalised and interconnected world represents optimal conditions for innovation. It was therefore inevitable that the stability of Paleolithic societies would give way to chronic instability, ideally oriented towards more power, more happiness, more wealth, more security, more innovations. Obviously, except for a major crisis alone capable of ending this dynamic, the pace of innovations can only accelerate: it is necessarily exponential and it is probably ridiculous currently compared to what it will be tomorrow. Today already, any human who would leave the planet for a few years would be amazed at the technological evolution upon his return. A few centuries ago, nothing really changed from one generation to the next. The conveyor belt of innovation is constantly accelerating, leaving little time for humans, little time for a brain that is still evolving at the slow pace of biology, to adapt to the new parameters of a constantly changing society. This is where Schumpeter's concept of creative destruction takes on its full meaning, as it is true that capitalism and innovation are intimately linked, this creative destruction being also inevitably one of the causes of the increasing precariousness that affects citizens at all levels of society.

If the Neolithic revolution provided the cultural environment conducive to an extraordinary exploitation of cognitive and creative human capacities, there is another psychological dimension inherent to human nature that has played and still plays a considerable role in our endless search for an ideal world. It is the awareness that we have of our incompleteness, an acute awareness of our vulnerability, of our

[12] Pringle (2013).

finiteness and, probably associated with all this, a background of metaphysical anxiety that religions have had the vocation to alleviate by promising us the possibility of eternal life and by proposing a general interpretation of the universe that makes sense and reassures us. There was a time when religion partly sufficed to alleviate this psychological discomfort inherent in the human condition. Our lives material and spiritual were in the hands of God, or the Gods, and any particle of incomprehension, mystery, psychic suffering, anxiety, frustration, was immediately erased by the religious or magical vision of the world.[13] The divine presence was enough to soothe the awareness of our limits and our incompleteness. It saturated our psyche as it saturated the world in its totality. But now, the sacred is hardly present either in our minds or in our environment, leaving us in direct contact with nonsense and our vulnerability as mortals. Men have developed two secular parades to overcome this fundamental psychological discomfort: infinite desire and the constant will to modify both one's being and one's environment through innovation and therefore technological progress. So there is at the origin of our frantic search for technical evolution, for transformation of our environment and ourselves, a form of fundamental dissatisfaction that can take various forms. It is not about pathological dissatisfaction—even if it can sometimes be—because it is constitutive of the human spirit. Men are clearly incapable of being satisfied with what they are or what they have. They are not psychologically in a state of homeostasis and are tragically incapable of achieving a stable psychic state. Where most animals consent to rest, to sleep and accommodate total inactivity, human will not survive without setting himself new goals and innovate to achieve them. Mental rest or physical rest is actually unbearable to us and we only accept and tolerate it for the most part after an episode of intense activity. We only find our balance through a form of chronic imbalance, in constant evolution and in the perspective of upcoming changes even if the price to pay is that of precarity, the risk of failure and that of transforming a planet to the point of making it unfit for our survival. This property of human psyche, undoubtedly the one that led human to brutal domination of his environment, is also the engine of our exceptional creativity, our frenzy of understanding of the universe and of course its mastery. A child of a few years will naturally wonder about the origin of our universe, about his own future once dead and about a quantity of natural phenomena that the most brilliant chimpanzee will peacefully ignore. We can say that the Neolithic revolution has allowed this psychic component of the human to express itself with redoubled force since it has instituted innovation and technological progress as a societal goal, as a means of domination over the world and over competing societies. We are now beings eager for changes, innovations, transformations of ourselves and our environment. Nothing would be more unbearable to us, with a few exceptions, than to live in a world that would no longer evolve and that would regain the boring stability of the world known by our ancestors in the Paleolithic. This stability would be frightening because it would no longer leave any hope for Men to alleviate the feeling of incompleteness that affects us all more or

[13] Barrett (2007); Bering (2006); Bloom (2004).

less strongly and more or less consciously. It is likely that we have deep within us the secret hope, in fact the naive illusion, that progress and technological innovation will bring us closer to an ideal state of well-being which in reality escapes us as soon as we believe we are approaching it. The illusion is all the greater because, in fact, technological innovation has brought us a lot and has undeniably made our daily lives more comfortable. Some voices have recently been raised to unmask this illusion and to point out the dangers that scientific progress and the appropriation of certain new technologies can represent. But the phenomenon is new. Again, however, the awareness of the dangers of certain technologies is hardly a match for the seduction they exert on us and the dependence they generate. To only take the example of 5G, how many among those who will have opposed it will consent to do without it once it is implemented? Here are a few tens of thousands of years that technological progress allows incredible improvements in our lifestyle. How not to succumb to the illusion that the happiness of men and their survival on this planet still depend on it? Very difficult not to give in to cornucopianism or even its extreme form transhumanism.

8.6 For a Regulation of Science and Not for Its Stop

There is an obvious fact: we would not be destroying our planet if scientific progress had not endowed us with a prodigiously efficient and powerful technology. It is indeed It's somewhat paradoxical to imagine that what was and is the cause of the environmental crisis today could be its remedy tomorrow. Yet this is a widely held and even theorised position by economist Simon Kuznets, a theory completely contradicted since, even today, the ecological impact of an inhabitant of a highly scientifically developed country is infinitely more significant than the impact of an inhabitant of a less scientifically developed country. Could it be that technological progress will be beneficial tomorrow when it was destructive yesterday, in terms of the environment of course? This is a very difficult question because the resolution of this problem is not technical or scientific. It depends solely on the ability of humans to reasonably and efficiently exploit the infinitely increasing power they will have over their universe: the resolution is therefore political in the broadest sense of the term and, as nothing is less predictable than future political orientations, any prediction here is more akin to astrology than serious forecasting.

The preceding section might lead the reader to infer that I am one of those people—whom President Macron associates with the Amish movement—who denounce the wanderings of science and perceive it as the deep and primary cause of the environmental crisis to the point of wishing for its partial or total hibernation. This is absolutely not the case. I am first and foremost a scientist by training, whether it be my investigations in the field of scientific ecology or in the study of cognitive and brain functioning. My attachment to the scientific approach is therefore strong, especially as I consider it a powerful antidote to a human irrationality that today represents a major risk for our societies.[14] The very idea of stopping our scientific investigations would seem to me to be the expression of a major regression and moreover perfectly unrealistic. As I have explained above, unless we imagine a totalitarian form of control or prohibition of scientific activity, we cannot put an end to what constitutes a fundamental and magnificent element of human psychology: our curiosity and our inexhaustible need to understand our universe. To arrive at such a solution would be a sign of a monumental failure of the human adventure, a brutal and regressive solution.

However, I must concede that I have moved in a few years from an unconditional admiration of science to a form of perplexity. The reason is actually quite simple. I think that our cognitive abilities have developed much faster than other components of our psyche, notably those that determine the mastery of our behaviours, the mastery of our relationship with others, the mastery of our desires, the mastery of our aggressiveness and more generally: the mastery of our relationship with the environment. In psychology, we admit the existence of functional and neural differences between the cognitive domain (memory, reasoning, learning, language, perception…) and the psychoaffective domain (which determines our emotions, our social behaviours, our empathic abilities, our affect…). Of course, these two domains interact strongly and many researchers are now studying their mode of interaction. The themes of emotion and memory, emotion and decision or emotion and reasoning are now classics in psychology and neuropsychology. Overall, and this has been remarkably explored by Kahneman and Tversky, we have extraordinary cognitive abilities. We have the ability to produce perfectly rational reasoning and decisions. This ability is probably recent on an evolutionary scale and is also unevenly present in each of us. But this ability is far from being able to express itself without being dramatically limited by more archaic forms of reasoning—older from an evolutionary point of view[15]—and by the largely unconscious pressure exerted from areas of the brain more associated with the psychoaffective sphere. The result is, for example, that a perfectly rational individual capable of very sophisticated reasoning in a given emotional context produces perfectly erroneous reasoning and makes incredibly irrational decisions in a different emotional context.[16] A similar phenomenon is observed in many so-called precocious children, or even some autistics, who have developed very high cognitive abilities for their age, but who remain very backward on the psychoaffective and relational level. In a certain way, the situation of *Homo Sapiens* today is identical. Our psychoaffective development (our ability to manage our emotions, our aggressiveness, our ability to reasonably project ourselves into the future, the management of our impulses, our selfishness, our fantasies of power and domination over others, our visceral narcissism…) is far behind our cognitive development, so much so that the cognitive sphere is too often enslaved to the psychoaffective sphere in its most archaic and dangerous aspects. A deplorable video

[14] Ripoll (2020).

[15] See the analysis of these two modes of cognitive processing presented in Chap. 4.

[16] Gervais and Norenzayan (2012).

circulated on the internet, in which we see a chimpanzee, to whom soldiers have just given a Kalashnikov, seize it and start shooting all over the place, making the soldiers flee. We then see him brandishing his weapon with pride as if he had suddenly become aware of the power that this weapon conferred on him. There was something pitiful and monstrous about it, but I couldn't help but see, in a metaphorical form, the situation in which men find themselves today, possessing a technology that they often use in a devastating way.

I would like to illustrate all this through an example that is quite extreme but ultimately very representative of the problem I have just to evoke. Undoubtedly, death and its preamble, ageing, constitute one of the most terrifying things we have to deal with. It represents the most brutal expression of our inability to definitively tame nature. The theme of immortality is at the heart of all religions and a large number of literary, philosophical or cinematographic productions. This is an area where fantasy sometimes substitutes for the most realistic considerations: before the flood, did men not already live several hundred years, 930 for Adam and 969 for Methuselah? A sign of a society that increasingly cannot bear the limits inherent in our human nature, medical and AI research is flourishing in the hope, if not of achieving immortality, at least of further delaying the onset of ageing and its crowning achievement: death. The position of transhumanists and a large number of libertarians is clear: "we must kill death". This would be the most beautiful expression of the general programme set by Descartes or Bacon in their time: to become master and possessor of nature. Everyone should be able to choose the moment of their death and there is no reason why it should be imposed on us. Men should be free to choose their destiny, a form of extreme existentialism that is supposed to make men Gods by giving them immense powers, including the ability to escape the most fundamental biological determinisms. In the libertarian universe of the GAFAM, the excitement is currently at its peak. There are countless Silicon Valley billionaires investing billions in start-ups whose goal is simply to defy death. "Google against death" is the subtitle of a "Time" edition where one of Google's founders, Sergey Brin, modestly states that: "We want to invest in something longer and more ambitious than usual." Laurent Alexandre, founder of Doctissimo, asserts "the imminence of a sublime victory of science" because "thanks to the lightning-fast progress of the life industry, the human who will live 1,000 years has already been born!" Others have already invested millions to be cryogenised. This is the case of Peter Thiel, one of the founders of Paypal who, worried about dying before a means of immortality has been found, has signed a contract with Alcor Cryonics, the world's number one in cryogenics, to be frozen after his death. Finally, others are betting on cybernetic immortality. Russian billionaire Dmitry Itskov, a media magnate, has committed to creating an avatar by 2045 in which the human being could duplicate his consciousness. He has even sent a letter to the thousand richest people on the planet to encourage them to finance his dream: transhumanism, associated with libertarianism, is clearly elitist! I don't know whether to smile or be exasperated by the arrogance of these individuals, their certainty, their desire to free themselves from natural determinisms and perhaps also their excesses. But one thing is certain. The frantic search for increased longevity is no longer the exclusive domain of wealthy and eccentric businessmen. Scientific publications on this theme are abundant and a number of recognised scientists predict the possibility of doubling or tripling the life expectancy in good health by the end of the century.

Without further analysis, one can immediately be struck by the part of *hubris* and madness that radiates from this field of scientific investigation. One can also grasp the dangers and the dreadful consequences that the realisation of such objectives could have on a societal as well as an ecological level. How could humanity, already cramped on our planet, survive if life expectancy were tripled? Yet, if one analyses the approach coldly, it is not so easy to categorically reject it. Why should we accept the limits of a nature that has no purpose, unless we adhere to a form of teleology that science has definitively rejected? Why should we accept without flinching the religious concept of imperfection or the scientific one of determinism? Why should we not do everything possible to perfect the human being to the point of subtracting him from death or ageing? Isn't there a form of basic and unreflective conservatism behind our desire to maintain the world as it is, a groundless timidity, or even a strange refusal to seek the best for human and to make him freer than he is? One of the prophets of transhumanism, David Nicholas, has a very concrete way of approaching the problem:

> It's very curious that people ask this question. No one wonders why Mother Teresa wants to save people who are dying in India! No one finds it abnormal that Louis Pasteur wanted to prevent people from suffering and dying from infection, he explains. So why do people ask me the question, who wants to save people from death? It's exactly the same!.[17]

This objection by David Nicholas calls for two remarks.

The first is the difficulty of positioning the cursor correctly, i.e. to what extent can and should science be developed with the aim of improving the human condition? The question is extremely complex as it is likely that there are multiple conceptions of what "improving the human condition" means. I would make a simple remark on this subject. If, by improving the human condition, we mean giving men a surplus of happiness then we are mistaken. All the studies conducted in psychology, already mentioned in Chap. 2, are clear and convergent. Individual happiness is not related to the improvement of our material living conditions, once these have reached a basic level that we have long since reached in most Western countries. In a number of cases, a surplus of comfort and material security is even liable to degrade our physical and mental state. The obesity epidemic, which affects 35% of American adults and 650 million people worldwide, is a dramatic illustration of this: the abundance of food is not only advantageous. The addiction to screens, which affects about 150 million Chinese people, is another. There is no doubt that 5G will contribute to its expansion. Of course, the alibi of individual freedom will be immediately reused to protect against such objections. Indeed, everyone is free to eat and connect moderately. But this freedom, as psychology has so clearly shown, is very theoretical, especially since the goal of economic growth at the heart

[17] Quoted in "A Dream Without End", a documentary broadcast by "Envoyé spécial", April 2015, directed by Anouk Burel.

of all national and entrepreneurial policies is doing everything possible to prevent the citizen from curbing his multiple addictions: growth above all, even at the cost of the physical and mental health of consumers. To return to the example of 5G, how will increasing download speed make people happier? How will being able to connect more quickly to the hundreds of connected objects in our environment increase our level of happiness? How will being informed a few seconds earlier about the outcome of a football match change your perception of the sport? Anyway, due to a habituation phenomenon that I have previously explained, as soon as 5G enters our lives, we will find that it is not fully satisfactory. We will then have to move on to 6G, which will meet the same fate and will not have any more effect on our individual happiness, but which our political leaders will support for obvious reasons of economic competitiveness. Human psychology is so archaic at times, so subject to bottomless desires, to the need to desire, so filled with a fundamental frustration, that it will be very difficult to spare ourselves the sterile and destructive exhaustion that the filling of our countless and pathetic Danaid barrels imposes on us.

If we return to the symptomatic case of life expectancy, and health in general, the difficulty of positioning the cursor is much more difficult than in the case of 5G and other countless new technologies. There is indeed no essential difference between the efforts of medicine to protect us from a virus or to treat us for cancer and those that transhumanists are developing to allow us to live much longer or even to escape death. To take just two examples among thousands, haven't doctors already produced an augmented human by developing vaccines that advantageously modify our immune defences or by installing a pacemaker to compensate for the defects of a tired heart? More concretely, what parent would refuse any form of medical progress if it could save their child suffering from a deadly disease? What wouldn't we accept when it comes to our health, or that of our loved ones, and the prospect of living longer and better? Who would refuse to live three hundred years under the same conditions that allow us to live 80 years today? Can we clearly position the cursor and decide on the basis of strong principles where scientific innovation can become dangerous? In reality, no fundamental argument allows us to decide because there is only a difference in degrees between the delirious objective of immortality and the more modest one of treating us when we are sick. The transhumanist argument, of the total freedom that we should grant ourselves in the aim of improving our existence, is no more convincing than the more conservative and religiously originated one, of the natural limits that should not be transgressed. In reality, any philosophical speculation on this theme consists in trying to find justifications for something extraordinarily basic and fundamental, but which we do not control. Innovation and creativity are simply the most radical weapon that evolution has provided us with to maximise the chances of survival of each individual and, of course, of the species. We are biologically compelled to always maximise our chances of survival and our chances of transmitting life, and many of our behaviours that appear so far removed from this fundamental objective are in reality subject to this requirement. This is very clear in the medical field precisely because medicine is the most direct way to stay alive and transmit it. This is the deep reason why we would accept any medical progress that could save us, even if it could have serious repercussions on humanity (this could be the case with a form of immortality). This is obviously the raison d'être of the national ethics advisory committee and the difficulties it has in ruling on many new technologies is very symptomatic. In reality, what appears quite clearly about health, is true of any technological innovation. This appears less crucial to us because these technologies do not directly affect humans, but the problem is absolutely identical. 5G is just a sophisticated form of what men have achieved by moving from chipped stone to polished stone. Innovation is only the expression of a fundamental and basic mechanism that leads us to want to indefinitely modify our environment to maximise our chances of survival. Once again, the whole problem is that what was adaptive in our prehistory and more recent history may no longer be so today, particularly because our planet is finite while our capacity for innovation is infinite. Associating the Amish ideology with those who have become aware of this problem is a grotesque way of ignoring a fundamental problem.

Those who believe that such a problem can be categorically resolved are greatly naive as the biological, and now cultural, forces that push us towards endless progress are powerful. As is often the case, extreme solutions, those that demonise science to the point of hoping for its end or those that rely on the utopian belief that its development, freed from all constraints, is the only way to solve "the complex problems of our society"—to quote our President—are perfectly disastrous and unrealistic. This is the meaning of the second remark we will make about the objection of transhumanist Daniel Nicholas. If it is neither impossible nor desirable to end scientific research, it cannot develop without safeguards because it is absolutely clear that the exponential nature of our technological power is potentially dangerous for humans as well as for our environment, and will become even more so tomorrow. Working in the field of cognitive sciences, I myself am confronted with a problematic state of cognitive dissonance. All the research we conduct in this field has obvious repercussions in the field of AI. Understanding how we understand, we remember, we learn… plays a fundamental role in the development of a superior Artificial Intelligence. The exploration of our cognitive abilities is fascinating and how could we be forbidden to want to understand who we are? In doing so, we know very well that our work in fundamental research will be very quickly exploited in the field of applications in order to design Artificial Intelligences that will, in the centuries to come, completely renew and disrupt our relationship to the world, our relationship to work, our relationship to others, our relationship to life and perhaps even the meaning of our lives. No other field than AI illustrates as well the fact that our intelligence is capable of conceiving something that surpasses us and of which we can therefore partially or totally lose control. The fundamental question that arises today is that of the possibility of evaluating the interests or potential risks arising from the development of a new technology. This is absolutely essential for both humans and our environment.

The task is in itself very difficult because, as we have seen, the boundary is thin between the acceptable and the unacceptable. Genetics is of immense theoretical interest and has led to dizzying medical progress. Recently, however, a Chinese researcher managed to immunise two viable babies from HIV by directly intervening on their DNA. Researchers around the world reacted with horror and fascination—both terms are important because they illustrate our ambivalence towards this type of technological prowess—because this could very well lead to fuelling the eugenic fantasy which is never very far away. The boundary is all the thinner because any legislative decision is always the result of a conceptual patchwork more or less theoretically supported since we do not have a priori criteria to distinguish the acceptable from the unacceptable in science. There is no other solution in this matter than to associate the competence of experts with political power in a constant concern for democratic demand. An ideal objective that we are far from achieving for various reasons that we will now analyse.

8.7 Democracy and Science

The first of these reasons lies in the fact that there is a growing gap between the speed of technological evolution and our ability to integrate it in a timely manner into a philosophical and political reflection space. In this frantic race, it is likely that politics will always be one step behind, firstly because our institutions have hardly changed in 50 years and are no longer adapted to the reactivity that the frantic pace of our innovations imposes, and secondly because many political leaders have no scientific or even philosophical training. This results in a major risk. Either politics shirks its responsibilities and leaves it to the experts to decide in their place: in this case, it is a clear denial of democracy. Or politics ignores the recommendations of the experts or ignores them, which is already evident for a large number of populist leaders. There is an urgent need to conceive today a structure equivalent to the national ethics advisory committee, but which would no longer be limited to research on humans and living beings alone.

The second of these reasons seems much more serious to us. It is the fact that public research, less and less supported by states, is giving way to private research. This is one of the inevitable consequences of the neoliberal evolution of our societies, of globalisation and deregulation as a principle. This results in an incredible loss of power for states. The friendly start-ups that these same states had clearly supported in the 2000s are now monstrous oligopolies with disproportionate financial powers that rival national powers, impose their own agenda on them and constrain them to levels still unimaginable two decades ago. The emergence of these monstrous companies with pharaonic budgets, notably the GAFAM but not only, has allowed their eccentric billionaires to invest billions in scientific projects that states could no longer support, among which, by the way, there are many projects dear to transhumanists. Undoubtedly, the launch of the Spaxe X rockets, on the very spot where NASA once propelled its own rockets, is the strongest symbol of this. Financial deregulation, is being followed by a deregulation of scientific research which, once in the hands of these gigantic companies, will escape more and more from any democratic form of control.

The third reason is linked to the previous one, but is nevertheless distinct. It is the inevitable entanglement of science and economy. A 2002 American study showed that investment in research and development between 1950 and 1993 had been responsible for 50% of the economic growth of countries such as the USA, the United Kingdom, Japan, Germany and France. Science is the main contributor to a nation's wealth and therefore to its geostrategic power, without ignoring of course that a significant part of scientific research directly contributes to the development of the national military arsenal, a key piece of the geopolitical weight of the richest states. Therefore, the problem is simple and for the moment without solutions. The primary concern of states is first economic success since it is this and only this that determines their hierarchical position in the world. But economic success is always relative to the success of other states. It is therefore a fierce economic competition that first determines political choices, well before any environmental concern. Therefore, the assessment of the interest of a new technology—5G for example—will always be made within each state according to the economic and geostrategic power that such an innovation can confer on them. Protective measures for the environment will only be adopted if they allow at the same time to ensure economic growth. As long as states maintain their dreadful and absurd competition, as long as unilateralism prevails, as long as the pursuit of economic and geopolitical hegemony remains their primary objective, as long as people, and especially political leaders, behave like archaic territorial animals, there will be no solution to this problem. No government will resist the additional power that any technological innovation could confer on it, whatever the environmental price to pay. For governments, the risk of delaying the exploitation of a new technology would be too great in the context of an economic war that requires always being ahead. From this point of view, moratoriums and the precautionary principle are perceived only as obstacles that states, like companies, will bypass as soon as they can. Only a representative supranational organisation could play a regulatory role in this matter. The issue being global and not national, no national structure could achieve the expected objective. But we must be realistic here. The archaism of human psychology, the strength of parochial altruism, the coarse and bellicose gesticulations of many political leaders, and the current rise of aggressive nationalisms, leave little hope for the possible emergence of such a structure and especially for the binding power it could have over states. It is enough to recall how President Trump sat on the Paris agreements to measure the archaic forces that currently prevail and which make almost impossible any real consideration of the environmental problem. Our national and international institutions, and therefore the structuring of the world into clearly distinct and always competing states—even when trade, cooperation or military agreements exist—is only a political consequence of elementary psychological forces that inevitably lead us to distinguish the in-group and the out-group, the friend and the enemy, the associate and the rival, the same and the other, the local and the foreigner. These forces can be considered as grotesque vestiges, regrettable residues of our evolutionary history. They remain nonetheless extremely prevalent as political men, naively or strategically, never cease to exploit them as it is easy to increase national cohesion and find an external cause to its own internal difficulties by stigmatising the other. The stratagem is as old as it is crude, but it still works wonderfully. Between favouring the economy of one's own country, and therefore the short-term satisfaction of one's voters, and favouring an entity as abstract as the planet at the risk of harming one's national economy, the choice is no mystery. It will remain the same as long as the negative effects linked to the destruction of the planet are perceived as minor compared to the disadvantages that could be linked to the adoption of binding environmental protection measures.

Beyond these obstacles, there remains the very difficult question of how to identify the criteria that such a transnational structure should adopt to assess the interest or danger of a new technology. Such a question goes well beyond the objective of this book. However, a very simple preliminary answer can be given. If we admit, in accordance with what any objective scientific analysis reveals, that to preserve the planet, there is no other solution than to stabilise our growth, then a new technology should only be accepted if it is likely to improve the lives of people without generating a surplus of growth and without of course harming our environment.

8.8 The Magical Conception of Ecology: Meeting Point of the Extreme Left and the Extreme Right

From a certain point of view, this intellectual movement, in its also very varied forms, is at the extreme opposite of the one I have just exposed. It is actually a quite old movement developed notably by the American precursors of ecology (Thoreau, Muir, Leopold…) and Europeans (Von Humbolt, Rousseau…), but whose most known and most important representative is the Norwegian philosopher Arne Naess. Naess is the founder of so-called deep ecology, a term he himself created in a famous 1973 article.[18] This ecology is said to be deep, as opposed to superficial ecology, the latter simply taking local measures to prevent environmental degradation from harming human development. On the contrary, deep ecology advocates a radical change in our way of life and in our relationship with nature. Naess therefore invites us to rethink our values, notably our illusion that the abundance of material goods, made possible by a forced exploitation of nature, will contribute to personal happiness:

> Ideological changes go through the appreciation of a good quality of life rather than adherence to ever higher living standards. We must become aware of the difference between "good quality" and "race to an extremely high standard of living" (which would be harmful to nature).

He thus insists on the need to reconnect with nature and to put an end to the relationship of domination that we exercise over it. This philosophy, even if it is often ignored by most activists and ecological activists, has played a considerable role in the representation that citizens make of what political ecology is. It has been the

[18] Naess (1973).

subject of a very harsh critique by Luc Ferry,[19] a partly founded critique, but which nonetheless consisted of caricaturing a revolutionary philosophy, extremely rational in many aspects, and far from the mystical, totalitarian, fundamentalist drifts it has nevertheless inspired.[20] In this matter, the greatest danger is indeed that of amalgamation. Deep ecology implies a form of Copernican revolution in the way we conceive our relationship to the world, a revolution that may mark a significant step in the evolution of human culture. As such, and because of the extent of the change in representation it implies, it was inevitable that it would provoke the most extreme and probably dangerous positions, as well as the most violent resistances from those who want to maintain the current political-economic system at all costs.

This philosophy radically breaks with anthropocentrism and with the belief that technical and scientific progress will provide a better life. The primary target of deep ecology remains, after all, the anthropocentrism of Western cultures, identified as the main cause of our devastating relationship with nature. This is the reason why Naess advocates a very Spinozist pantheism that has nothing mystical, which escapes any form of transcendence and which refuses the existence of a strong separation between human and nature. Human is an element of nature and his personal fulfilment depends on his ability to live in good harmony with it. In the platform of deep ecology, it is explicitly stated that "the well-being and flourishing of human and non-human life forms on Earth have intrinsic value, independent of the utility that humans might give them". The hierarchical relationship we have established between humans and nature must be definitively abandoned. Animal communities are the equal of human communities and we have a moral duty to respect the former as much as we respect the latter:

> Human beings do not have the right to reduce biological richness and diversity, except to meet vital human needs.

In the case where human activity would contribute to a degradation of nature, it would be up to men to accept a reduction in its growth and, if necessary, a reduction in its population:

> The flourishing of human life and cultures is compatible with a substantial decrease in the human population. The development of non-human life forms requires such a decrease.

It is not insignificant that David Thoreau (1817–1862) and John Muir[21] (1838–1914) have strongly inspired Naess's philosophy. Both had a scientific, romantic, aesthetic, poetic and almost mystical approach to nature. The starting point of their sensitivity to nature was triggered by the emotion felt in its presence and by the observation that many remarkable natural spaces have been ravaged by the increasing exploitation that men have made of them in their frantic search for profit and improvement of their material life. The primary motive for their commitment was

[19] Ferry (1992).

[20] Flipo (2010).

[21] John Muir was the originator of the "Sierra club", one of the most important environmental organisations and the creation of the world's first natural park: Yosemite Park.

therefore not so much to save humanity from the environmental chaos of which it is the cause, but to save nature, for its beauty and for the sacred, mystical and almost religious relationship we can establish with it. For Muir in particular, nothing can replace the direct experience of wild nature, which leads to the revelation of its sacred character from which Men have tragically cut themselves off due to the artificial and urban environments in which they now live. Not only does nature have intrinsic value, and not a value by virtue of what it can produce for Man, but its sacredness requires us to protect it whatever the cost and even if this requires limiting the power we exercise over it. From this point of view, science, if it is not bad in itself, would nevertheless be at the origin of our technology and its terrible impact on nature. Even if Naess expressed less clearly than his predecessors a form of mystical attachment to nature, even if he remained a humanist who refused to place nature above Man, even if he did not display explicit defiance towards science and attached essential value to rationality, his work has inspired a mystical-magical ecological movement, sometimes anti-scientific, anti-humanist, sometimes violent and, in most cases, probably too radical to have a profound effect on the mentalities of our political and intellectual leaders.

An entire current stemming from the counter-culture of the 60s in the USA was inspired by Naess's work. It features in an extreme form the main lines of deep ecology to which Naess would absolutely not subscribe. The principle of equality between different animal species, antispécisme, is prominent and has provided a theoretical basis for the most determined ecological activists. The "Earth First" movement is a good example and one of their most famous slogans testifies to their radicality and the magical foundation that is theirs: "no compromise in the defence of Mother Earth!". In an interview with a journalist who asked one of its founders about measures to eradicate world hunger, the response was this:

> The worst thing we could do in Ethiopia would be to help the destitute. The best thing would be to let nature find its own balance, to let the people there starve to death.

The response caused a scandal. It nevertheless reveals a very real component of radical and misanthropic ecological movements, sometimes tinged with racism, which ends up idolising Mother Earth in a form of magical paganism as irrational as it is dangerous. The very term "Mother Earth" is a perfect illustration of the magical conception developed here. Bron Taylor, a specialist in religions, has called this current "black and green religions",[22] characterised by a rejection of traditional religions and an adherence to an animism perceived fantastically as the most respectful religion of the environment. Very typical of the most naive magical thinking, the Earth is perceived as a thinking entity, endowed with feelings and intentions and the power to make men pay for the mistreatment they exert on their environment. Many magical interpretations of this type have indeed been evoked to explain the Covid 19 pandemic. In fact, a very spiritualist ecology is emerging here, driven by the sacred duty to protect the Earth, an Earth that is more valuable than humanity itself. These ecologists are often millenarian and reconnect with an apocalyptic vision

[22] Translation of "dark green religion". See notably Taylon (2010).

stemming from Christianity which they nevertheless reject. This ecological radicalism frequently anti-humanist can go as far as advocating the disappearance of the human species. This is the case of the "Voluntary Human Extinction Movement" which sees no other solutions to protect the environment than to work for the extinction of the human species, the members of this movement indeed encouraging humans to stop reproducing to gradually make our species disappear. This radical approach has often radiated to more serious environmental defence organisations and have been the cause of distressing conflicts with local populations who denounced the attitude of wealthy foreigners, despising the fate of local populations and having in their eyes less value than certain threatened species: unfortunate legacy of a form of contemporary colonialism clumsily leading non-governmental organisations to transpose abroad the model of American national parks.

In a way that is not so strange, this extreme conception of ecology is also claimed by a far-right anti-humanism. This was the case with the neo-Nazi Savitri Devi, a pioneer of far-right ecology. It is an ecology imbued with social Darwinism, racism, Malthusianism, neo-paganism and a magical conception of nature dear to a far right which, through its adherence to a mysticism and esotericism of pagan inspiration, manages to free itself from the constraining rationality of the Enlightenment. There is therefore a far-right ecological activism, anti-technicist, anti-progressive, anti-rationalist, obscurantist which draws inspiration from deep ecology even though, tragic irony of history, Naess was a fierce and committed opponent of the far right and Nazism. Behind the ecological facade, behind a form of romantic sentimentality towards the animal cause and a fantasised return to a pristine and perfect nature, there is of course the xenophobic soil, the most violent and anti-democratic identity claims. A surprising convergence is emerging today between the discourse of the National Rally and a part of the left, notably through bio-regionalism and localism. It has become the basic ecological slogan of the RN.[23] Fight globalism, favour, the local, the terroir, tradition and escape international finance. The subterfuge is crude as it is easy to identify that it is only the visible and sympathetic face of a hidden identity obsession, xenophobic and clearly anti-humanist, the only one that really matters for all national far rights.

There is also, in the political forms that claim to be deep ecology, a totalitarian temptation that opponents of ecology are constantly waving today as a perfect deterrent for those who could be won over to the environmental cause. The fundamental question here is whether there is a necessary link, any theoretical convergence between ecology and totalitarianism. This is a question that deserves a slightly more subtle analysis than the one we are usually given. Let us simply note here that if the history of the twentieth century reveals possible affinities between fascism, totalitarianism and ecology, this does not absolutely allow to legitimise the hypothesis of a strong relationship between these currents. Such an inference is unacceptable and associating, without further analysis, ecology and totalitarianism is a refrain that has

[23] The RN (Rassemblement National) is a French far-right political party.

tough skin to take up an expression from Dominique Bourg.[24] Let us also note that, with the exception of extremist currents as found in just about all radical ideologies, most current ecologists assert a strong democratic requirement. However, it must be admitted that an important element of deep ecology, an element that has indeed been exacerbated by those who claim the philosophy of Naess, can contribute to a form of totalitarianism. It is antispécism. In itself, speciesism is probably more difficult to support theoretically than antispeciesism. However, any philosophy that claims to be able to rationally legitimise one or the other of these positions is misguided.[25] Unless the religious argument is brandished, that of a divine work that would have granted more values to Men than to the rest of nature, there is no rational argument allowing to support a hierarchy between human and nature. In reality, the superior value that we give to our species comes from biological forces as fundamental as they are archaic, those that make that any other animal species, yet perfectly deprived of the possibility of philosophising, will systematically favour a member of its own species over a member of another species. There is a form of specific chauvinism, biologically and culturally determined, which is only an adaptive mechanism among others to maximise the chances of survival of its own species. It is the same mechanism that makes that if, driving your car, you must abruptly choose between crushing a rabbit and crushing a child, you will crush the rabbit without even having thought about the legitimacy of this decision. We can put the philosophical varnish that we want behind all this, it will still be only a varnish, not a rational legitimisation. The antispeciesists try to break with this natural determination and it must be recognised that from a strictly ecological or biological point of view, human has no more value than any other animal species. It is undoubtedly from this questioning of a human chauvinism that imposes itself on us that the totalitarian risk comes from and this for two different reasons. The first is that any ideology that takes the counterpoint of what our most instinctive psychology pushes us to do will only be able to achieve its ends by imposing "its law" in an authoritarian manner. Our chauvinism is so powerfully inscribed in our genes that culture can only provide a counterweight by being imposed in an authoritarian manner. The second is that from the moment when the speciesist hierarchy is abandoned, or even reversed for some extremist militants of deep ecology, the fate of humanity can become secondary to the requirement of environmental protection, hence some antihumanist positions advocating an authoritarian regulation of human behaviour. Again however, we must not throw the baby out with the bathwater. Antispeciesism marks a cultural evolution that has allowed us to moderate a human chauvinism that is undoubtedly excessive. The fact that it leads us to respect other animal species more is a significant progress in our relationship with the living. It is our responsibility to integrate it into our conception of nature without it leading to extreme forms, eminently questionable and dangerous.

[24] Bourg (2019).

[25] Colyvan et al. (2009).

References

Barrett, J.-L. (2007). Cognitive science of religion: What is it and why is it? *Religion Compass, 1*(6), 768–786.

Bering, J. (2006). The folk psychology of souls. *Behavioural and Brain Sciences, 29*, 453–498.

Bjørn, A., Kalbar, P., Elsborg Nygaard, S., Kabins, S., Jensen, C. L., Birkved, M., Schmidt, J., & Hauschild, M. Z. (2018). Pursuing necessary reductions in embedded GHG emissions of developed nations: Will efficiency improvements and changes in consumption get us there? *Global Environmental Change, 52*, 314–324.

Bloom, P. (2004). *Descartes' baby*. Basic Books.

Bourg, D. (2019). Is ecology authoritarian, even fascist? Or decidedly democratic? *Ecological Thought, 3*(2), 1–13.

Colyvan, M., Linquist, S., Grey, W., Griffiths, P., Odenbaugh, J., & Possingham, H. P. (2009). Philosophical issues in ecology: Recent trends and future directions. *Ecology and Society, 14*(2), 22.

Costanza, R., Kubiszewski, I., Giovannini, E., Lovins, H., McGlade, J., Pickett, K. E., Ragnarsdóttir, K. V., Roberts, D., De Vogli, R., & Wilkinson, R. (2014). Time to leave GDP behind. *Nature, 505*, 283–285.

Costanza, R., McGlade, J., Lovins, H., & Kubiszewski, I. (2014). An overarching goal for the UN sustainable development goals. *Solutions, 5*(4), 13.

Drews, S., & van den Bergh, J. C. J. M. (2017). Scientists' views on economic growth versus the environment: A questionnaire survey among economists and non-economists. *Global Environmental Change, 46*, 88–103.

Ferry, L. (1992). *The new ecological order*. Grasset.

Feygina, I. (2013). Social justice and the human–environment relationship: Common systemic, ideological, and psychological roots and processes. *Social Justice Research, 26*, 363–381.

Flipo, F. (2010). *Deep ecology, a threatening fundamentalism or a non-modern liberalism?* Cahiers Sens Public Assoc. Sens-Public.

Gervais, W. M., & Norenzayan, A. (2012). Analytic thinking promotes religious disbelief. *Science, 336*, 493–496.

Jonas, H. (1979). *The principle of responsibility, an ethic for technological civilisation*. Cerf.

Kahneman, D. (2011). *Thinking, fast and slow*. Penguin.

Kallis, G. (2011). In defence of degrowth. *Ecological Economics, 70*, 873–880.

Metzner, R. (1994). The emerging ecological worldview. In M. E. Tucker & J. A. Grim (Eds.), *Worldviews and ecology: Religion, philosophy, and the environment* (pp. 163–172). Orbis Books/Maryknoll.

Naess, A. (1973). The shallow and the deep, long range ecology movements. *Inquiry, 16*, 95–100.

Pringle, H. (2013). The origins of creativity. *Scientific American, 308*(3), 37–43.

Ripoll, T. (2020). *Pourquoi croit-on ? Psychologie des croyances*. Auxerre, Sciences Humaines.

Sajaloli, B., & Grésillon, E. (2016). The Catholic Church, ecology and the protection of the environment: chronicle of a theological and political conversion. *Géoconfluences* (posted online on 19 October 2016).

Sanchez, P. (2007). *The rationality of magical beliefs*. Droze.

Taylon, B. (2010). *Dark green religion: Nature spirituality and the planetary future*. University of California Press.

White, L. J. (1967). The historical roots of our ecologic crisis. *Science, 155*, 1203–1207.

Zarzan, J. (2005). *Against civilization: Readings and reflections*. Feral House.

Chapter 9
The Human Trajectory: As Extraordinary as Despairing, but Contingent

Abstract In the previous chapters, I have emphasized the powerful biological and societal determinisms that lead us to this paradoxical situation where awareness of the impending ecological disaster does not lead us to change our individual and collective behaviors. I wanted to show that we should not, however, conclude that this is a form of inevitability. I analyze the strong relationships between competition, inequality, and the destruction of the planet while showing that this disastrous triptych can be overcome. If it was put in place during the Neolithic revolution, humans are cognitively flexible enough to free themselves from the determinisms that have led them to the current crisis.

If we try to take stock of the trajectory followed by our species since it appeared on Earth, I believe we can reasonably say that it is mixed. Again, the conceptions to the Cassandra or the Candide, have little interest and can only lead to inaction. Many of those who lament the living conditions in our society would benefit from making a short stay in the centuries that have preceded and those who are blissfully satisfied with the current situation would undoubtedly quickly revise their position if they were forced, even for a few days, to live as some billions of poor and destitute humans live in 2021. We have gained in many areas in just three centuries: politics, culture, democracy, freedom of expression, rationalism, standard of living, health and life expectancy, food security, everyday security, leisure, human rights… If I had to choose the time when I would like to be born, I would undoubtedly choose our own and I think that if we were capable of drawing up a faithful picture of our living conditions throughout our history, most of us would do the same. Steven Pinker[1] has remarkably shown this through a very fine comparative analysis of our living conditions throughout our history. Reminding this kind of evidence, which

[1] Pinker (2012).

T. Ripoll, *Human Psychology and the Destruction of the Planet*, Anthropocene – Humanities and Social Sciences, https://doi.org/10.1007/978-3-032-02080-2_9

the media do not relay enough, is very important because a negatively biased perception of our recent history serves as an alibi for anti-humanist, populist, nationalist, anti-democratic political currents that wish to put an end to the rationalism of the Enlightenment on the pretext that it is the cause of all our ills.

If the balance is partly positive, it nevertheless suffers from three major defects that could alone put a painful end to our trajectory and plunge us back into a barbarism, a violence and obscurantism that are just waiting for major crises to reactivate. The important point I would like to expose is that these three major defects are intimately linked psychologically as politically. They have the same origin and apprehending them together is necessary to solve the problems we are facing. These three defects concern the level of inequality, international competition and the resulting violence, and of course, the destruction of our environment. The environmental problem will have no solution as long as the level of inequality remains that of today and as long as competition remains at the heart of our relations at the individual, national and international level.

The primary cause of the ecological disaster is the objective of continuous growth to which all states, and most individuals, aspire. The pursuit of this objective makes no sense for two obvious reasons reasons. The first, psychological, is that the level of well-being is decoupled from economic growth as soon as GDP reaches a value that we have reached for over 50 years in the west.[2] It even seems that, associated with a increasing level of inequality, the well-being of citizens of rich countries has declined over the last 20 years. The second reason is unquestionable on a theoretical as well as empirical level. Whatever the level of technological perfection, economic growth inevitably causes environmental degradation which are also increasing, simply because the gain in technological efficiency cannot compensate for the effect of the increase in GDP on the environment. Decoupling, the cornerstone and magic of the cornucopian ideology, will unfortunately never be sufficient to ensure that economic growth eventually ceases to impact the planet at all. On the scale of growth that should continue over a long period of several centuries or millennia, the cornucopian belief is a worrying utopia or simply a form of absurd magical thinking[3] consisting of taking one's dreams for reality, like the famous phrase of President Reagan, champion of unbridled capitalism and notorious anti-environmentalist: "There are no limits to growth when men and women are free to pursue their dreams".

In the current economic system, we are growing both too fast and too slow. Too fast to preserve the planet from the effects of our production/consumption, too slow to ensure full employment. The consequence of all this is crystal clear. It is not possible to indefinitely maintain our growth objectives. Assuming that we stop growing, the only possible alternative if we want to overcome the environmental crisis, the whole question is whether it is possible to preserve our level of well-being and our standard of living given the level of global wealth that we can produce today.

[2] Kubiszewski et al. (2013).

[3] Kallis et al. (2018).

9.1 The Three Faces of the Same Reality: Competition, Inequality and Environmental Destruction

Intraspecific competition is a fundamental element of life. It plays an important role in the evolutionary process by allowing individuals best adapted to their environment to impose themselves on others, to increase their opportunities for reproduction and thus to optimise the probability that their genes survive their death. Humans are not exempt from this fundamental dimension of life. Of course, the aggression linked to any form of competition can be dangerous for rival individuals, or for the species, if it is not regulated, either by biologically determined processes themselves, or by rules specific to each culture. The survival of all societies of social animals depends on very complex processes allowing competition to be exercised optimally, altruism and cooperation generally constituting an effective counterweight to the risk inherent in any form of unregulated rivalry. As far as our species is concerned, all anthropological research has shown that before the Neolithic, violence resulting from competition was already present, probably more between distinct groups than within a group for the simple reason that an excess of violence in a small group would have endangered its survival. Within a group of Paleolithic humans, let us remember that inequality was strongly rejected, which is why competition probably could not have the importance it has in cultures accepting inequality: why compete and develop a rivalry relationship if in the end the winner has nothing more to gain than the loser? The Neolithic revolution completely reshuffled the cards. For the first time in human history, some people became rich while others became poor. The Neolithic invented the class of beggars and the class of exploiters: it is from this time that slavery dates.[4] Inequality quickly became a norm that we have all integrated regardless of our aspiration to social justice and our natural aversion to inequality. Men were then inevitably put in competition at a time when institutions were probably not yet organised enough to regulate it. It is quite "amusing" to note that neoliberalism is trying by all means to reconnect with this old deregulation. This resulted in great violence and many anthropologists and psychologists[5] of religions consider that men invented Gods so that, from their incontestable and transcendent voice, they managed to impose on men the first instituted rules necessary for the peaceful life of these new human communities: the table of the ten commandments is an example.

It is undeniable that the disappearance of the constraint of equality and the concomitant possibility of the freedom to enrich oneself have led men individually, then collectively, to initiate an ever-increasing cycle of innovations. Competition now structures the lives of men and, by extension of groups of men, as they quickly constituted themselves into cities, regions, empires and states. The conflicts that once opposed modest tribes of a few hundred men quickly opposed empires of millions of individuals. As the geopolitical power of these empires was directly dependent on

[4] Pathou-Mathis (1993).

[5] Pyysiainen and Hauser (2010).

their wealth and this latter is itself dependent on the natural resources of their territory and the ability to exploit them, it became essential to progress technologically. As far back as we can go in history, technological advance has been the most effective weapon allowing the most ingenious to impose their domination, whether directly through the development of new military technologies or by better exploitation of natural resources. Things have not changed much today except for the sophistication of our technology and our increased ability to exploit nature to the point that our survival on this planet seems compromised. Competition, inequality, destruction of the planet: this is a powerful and devastating triptych that can only be mastered if its three components are treated together.

9.2 Competition Is a Major Obstacle to the Preservation of the Planet

Inequality as a cultural norm and competition as a driving force of this inequality has undoubtedly played a major role in the trajectory followed by humanity over the last 10,000 years. Capitalism and neoliberalism are not mistaken: curbing competition and rivalry risks slowing our dynamism, creativity and growth. Indeed, on a government website, there is clear adherence to this ideology through the explanation given for France's low growth rate in recent years:

> The explanation lies mainly in the greater rigidities in the organisation and structure of its labour market and the excess of regulations that have slowed competition.[6]

To perpetuate the system and make it even more dynamic, to continue to grow, the least possible regulation is the best solution, such is the refrain of liberalism. The problem is that this logic is now fully integrated by all states to the point that each state, in a definitively globalised world, is trapped by the ideology it has adopted willingly or by force. The system, in the cybernetic sense of the term, has taken precedence over individuals and their ability to renounce the rules that govern it. It is enough for a nation to lose its competitiveness for its trade balance to become deficit and for its economic balance to darken. This will inevitably result in a slowdown in growth with its disastrous consequences in terms of employment and therefore social crisis. In short, states have no other solution than to play the deadly game of who will reach the highest level of growth regardless of the environmental and social consequences. To do this, they will naturally go into debt to invest because there is no growth without investment. But as they will then have to repay their debts, they will have to grow even more because only the budget surplus provided by growth allows both to reduce the debt and to provide the minimum social aid without which the poorest citizens could revolt. The rating agencies, the ultimate constraint invented by the system, are there to remind states of the need to grow

[6] https://www.economy.gov.uk/easyeco/growth

because if growth is not there, the rating they will give them will decrease which will lead to an increase in interest rates and therefore their level of debt. States then have no other solution than to adopt austerity plans that will burden the poorest, to reduce corporate taxes and to restrict public spending in the hope of a final surge in growth. But the era of "easy" growth has probably come to an end. Mature economies have now entered the era of stagflation and then stagnation. Since the golden age of the 60s, growth has been steadily decreasing worldwide even though it has been maintained at a moderate level thanks to the rise of emerging countries. This slowdown in growth is not cyclical: it is structural. Even technological innovation, planned obsolescence, large-scale waste, increased productivity at work, the excesses of fantastic advertising campaigns... cannot counteract a heavy trend linked to the slowdown in world demography, the fact that consumption has reached a plateau and the fact that the price of energy will naturally increase. But the capitalist system is a system that collapses if it does not grow. There is something truly pathetic and absurd about the extreme measures that have been adopted, in an urgency that smacks of panic, to maintain a minimal level of growth necessary for the survival of the system: disappearance of national borders for free movement of products and capital, states stepping back in their regulatory mission, large-scale privatisation of public companies, lowering of taxes, regression of agreements with social partners, lowering of wages to increase competitiveness, and even increasing working hours... Freeing the brakes of economic growth or dying is the alternative to which all states are subject. Not only is there a fundamental incompatibility between such a system and the resolution of the environmental crisis—except to indulge in the cornucopian utopia of green growth—but no state can escape the system because any measure aimed at protecting our environment that would not lead to an increase in growth is a measure that states cannot adopt, except to commit economic suicide. This is why the French state has recently invested heavily in housing renovation to minimise energy expenditure and has refused, at the same time, to tax French civil aviation even though both measures have clear ecological relevance. In the first case, housing renovation is part of the context of an ecological transition that has the double advantage of reducing our energy consumption while guaranteeing an overall increase in growth. It is economically ideal and perfectly compatible with the programme of green growth. But we must not be deceived. The inevitable rebound effect will be such that the ecological gain obtained by renovating housing will inevitably be offset by an increase in consumption. To give just one well-documented example, the saving made by a household in its heating budget is immediately used, either to heat more (direct rebound effect), or to benefit from other types of consumption due to the saving made thanks to better thermal insulation (indirect rebound effect). Growth is only green in appearance because a global balance would easily show that the gain in technological efficiency has produced a global increase in consumption with its inevitable impact on the environment. In the case of taxing civil aviation, the pattern is very different. This type of taxation does not directly generate any growth. It even has the double disadvantage of reducing growth and reducing the competitiveness of the company. It reduces growth since the increase in the price of the plane ticket tends to reduce air traffic and it reduces the competitiveness of the company because it increases the cost of its services. Company competitiveness since consumers will naturally turn towards competing companies offering lower prices. The dilemma is simple: either we tax polluting activities and reduce the competitiveness of the companies concerned, with the risk of social breakdown; or we take no environmentally virtuous restrictive measures in order not to affect the competitiveness and growth of national companies, and these companies will do nothing to reduce their ecological impact as long as competing companies do not do the same. We can analyse it as we wish, we can develop sophisticated rhetoric to circumvent reality, but competition between companies and between states is probably the first of the obstacles to resolving the environmental crisis. Any environmental protection measure likely to impact the competitiveness of companies and reduce national growth will always be sacrificed or deferred for reasons that have real short-term economic rationality and profound long-term irrationality. There is a fundamental incompatibility between liberalism in economics and serious consideration of environmental problems.[7] The fundamental reason for this is very simple. In competition between states, growth is the lethal weapon, lethal for imposing one's domination on others. The fable of green growth to which some adhere out of pure ideology or ignorance of the scientific reality of the environmental problem is the crudest stratagem that could have been invented to convince ourselves that the system could continue in a context of environmental crisis.[8]

9.3 Can We Live Without Competition?

In theory, yes. Nothing prevents us from interacting with others without the interaction being contaminated by some form of competition. The reality is much less simple. Humans are beings of desire. And as we often have the same desires and it is not always possible to satisfy them all, antagonistic relationships almost inevitably appear. Moreover, we often complain about excessive competition and yet part of us continues to cherish it. If this were not the case, how would we explain the joyful excitement that takes hold of millions of humans on the occasion of major sporting events? Why do we love to challenge ourselves or simply play with friendly opponents in many situations in our social life? We are so psychologically dependent on competition that we prefer the risk of losing, if the chance of winning is not nil, to accepting an outcome for which there would be neither loser nor winner. Boredom is our greatest terror, even more so than that of failure. What is true at the individual level, and which the Neolithic revolution exacerbated due to its acceptance of inequality, is naturally found in the competition between states. There has never been in the history of humanity, the slightest parcel of time that has escaped war or extreme international tensions. When we see today, at a time when more than

[7] Kallis et al. (2018).

[8] Bjørn et al. (2018).

ever states should coordinate and cooperate to face the environmental crisis, how much leaders play the national, expansionist and xenophobic card and how much they manage to be elected on the basis of nationalist slogans, we can only be afflicted. Competition was undoubtedly one of the essential drivers of the dynamic that led human to create the sophisticated society we know today. But this same engine, once so productive and energising, now threatens us very directly. It threatens us through the inherent risks of major military conflicts, now capable of destroying the planet given the power of the weapons we have. It threatens us above all through this fierce competition that states wage to increase their national growth more than that of their neighbours, a competition dramatically incompatible with the need for a global treatment of the environmental crisis. As long as men do not manage to moderate their aggressive impulses, as long as they remain driven by archaic forces that lead them to consider the other as a rival or an adversary (at the individual level as well as at the international level), as long as a desolate form of parochial altruism will force them to favour one part of humanity at the expense of another part, it will be impossible to adopt effective environmental measures. What is certain is that in the fragile balance between competition and cooperation, selfishness and altruism, individualism and solidarity, our society has favoured and encouraged the first of these two terms. There is nothing natural or inevitable about this since other human cultures have positioned the cursor differently.

9.4 Inequality as Another Major Obstacle to Preserving the Planet

If competition creates inequality, the latter fuels competition, both constituting the two sides of the same and unique phenomenon. Today, income or wealth inequality has reached historical levels within nations and between nations.[9] Far from decreasing with the increase in global wealth as thought[10] or still think liberal economists, it is accentuated on the contrary with the level of global wealth. Historically, the gap between nations has amplified dramatically from the sixteenth century. This is a shift that has been called the great divergence and which corresponds to the rapid development and enrichment of Western economies, triggered by technological progress, a new social and political organisation and the appropriation of resources from the first colonised lands. Before this time, it should be kept in mind that between an average Chinese, African, European or Native American individual, the level of wealth was very similar.[11] For two centuries, the increase in inequalities

[9] Piketty (2013).

[10] See the supposed Kuznet's law.

[11] Bourguignon (2012).

between countries has continued to increase at an accelerated pace so much so that between 1820 and 2002, the Gini coefficient has almost doubled.[12]

The level of inequality is increasing globally in the world and has reached a level that nothing can justify philosophically and ethically except the sacrosanct law of competition and competition. On the level of social cohesion, as well as on the level of mental health, inequality has a catastrophic effect[13] well documented. The more unequal countries are, the more social and psychological distress increases, the more the health of citizens as a whole deteriorates even if it deteriorates more strongly in the disadvantaged social classes. This is particularly the case in the USA where obesity, diabetes, drug addiction, multiple forms of addiction, social phobia... strikingly affect the poorest individuals. It is still in the USA, the richest country in the world but very unequal, that the rate of depression is tragically high because if the poor suffer from their status, the middle class lives in chronic fear of downgrading. According to the latest report on mental health in the USA, 70% of young people have experienced a major episode of depression and 56% of adults suffering from depression are not followed. All these deleterious effects of senseless inequality explain that the increase in wealth that Western countries have experienced in recent decades has not contributed in any way to the increase in individual happiness. The quest for well-being is unfortunately undermined by the scourge of inequality which remains what men have invented worst to break cohesion and social ties which are powerful contributors to happiness.

But inequality is also a major obstacle to the adoption of ecological measures for two fundamental reasons. The first is that the level of awareness of the environmental crisis is a function of the level of education. However, income inequality is tragically correlated with inequality in terms of education. A large part of humanity remains insensitive to the environmental crisis simply because it does not have the necessary scientific information or it is not able to process them effectively. The second reason is probably even more important. Inequality has a very clear translation in terms of consumption and therefore environmental impact. If you wonder if your individual environmental impact or your level of emission in terms of CO_2 is more or less important than that of the average world population, the answer is very simple with a few statistical nuances. Participation in the destruction of the planet is simply determined by the level of income and wealth. A study[14] showed that when a woman gives birth to a child, one can estimate from the mother's standard of living, the amount of additional CO_2 that will result from this new individual. If we compare an American mother and a Bengali mother, the carbon production emitted as a result of this birth will be approximately 130 times greater for the American mother. The richer you are, the more you pollute: the 1% richest of the planet emit twice as much CO_2 than the poorest half of the planet. The same is true if we measure the ecological footprint which is insignificant in a poor country like Niger and

[12] Milanovic (2007).

[13] Pickett and Wilkinson (2019).

[14] Murtaugh and Schlax (2009).

reaches peaks in a rich country like Qatar. For a given country, the relationship between wealth and environmental impact is strictly identical. Jean Gadrey estimated that the richest French people pollute 40 times more than the poorest.

The consequences of all this are absolutely disastrous at the national and international level. What relatively poor or even moderately rich country would agree to reduce its consumption and its production of wealth when others contribute much more than them to the destruction of the planet? How can we demand from the poorest countries that they adopt restrictive ecological measures when the citizens of rich countries continue to have an infinitely larger ecological footprint? In this endless race for wealth, growth and consumption, no country will agree to moderate as long as the wealth gap remains what it is today. The same is true at the national level and this impacts the ordinary behaviour of millions of French people. It is absurd to ask for any environmental effort from modest French people when the richest ostentatiously display an incredible debauchery of consumption. And since there is always someone richer and more consumerist than oneself, inequality provides a perfectly understandable alibi for any reasoned modification of one's own behaviour. "I'm not going to deprive myself of travelling by plane from time to time when others drive Ferraris and travel every winter to the tropics" was the argument that one of my naturalist and ecologist friends returned to me when I questioned him about his decision to take the plane for a trip of a few days. The "yellow vests" episode and previously that of the "red caps" provide a remarkable illustration of the impossibility of taking reasonable environmental measures in a context of inequality and social insecurity. In both cases, it was a matter of introducing an eco-tax, the ecological and health legitimacy of which cannot be denied. But how could modest people accept a tax that would financially weigh on their budget to the point of making their "end of the month" difficult when others have such exorbitant assets and incomes? We must be realistic, at the individual level, as at the collective level, the adoption of ecologically reasonable behaviours often implies a renunciation or a reduction of one's standard of living: renouncing to take the plane, accepting to live at a temperature of 19 degrees, not air-conditioning one's home or, even worse, one's car, turning off one's mobile phone as soon as possible, consuming less goods from the other end of the world, renouncing to have a swimming pool built, renouncing to one's second home, travelling less, not saturating the clouds with thousands of photos that one never looks at, renouncing to have a dog or a cat… The almost infinite list of these small pleasures that seem like nothing but which, taken together, determine a level of consumption that the planet could not support if it was generalised to almost 8 billion humans who aspire to these same pleasures. This renunciation is psychologically difficult for all the reasons mentioned above. It is only possible to consent to it from the moment when citizens have a clear awareness of the close link that exists between their consumer behaviour and the ecological disaster of which they may be the first victims before their descendants are the certain victims. But even when mature and responsible citizens accept this reality, it is unlikely that they will agree to change their behaviours as long as others, infinitely richer, continue to live in ostentatious luxury. Competition, and the inequality that necessarily results from it, constitute insurmountable obstacles to any ecological revolution, which is why there is a fundamental incompatibility between capitalism, neoliberalism and ecology.

9.5 The Triptych Competition, Inequality, Environmental Destruction: An Empirically Confirmed Psychological Reality

In our culture, the word "competition" or "competitive" are now very positively connoted words. The call for competitiveness is commonplace within public organisations, companies, states and, on its altar, no one disputes the necessity to sacrifice human life, whether it be their quality of life, their salary, their working time, the quality of their social ties… and naturally, the planet. Competitiveness is a new norm that we have all integrated and to which we must conform to save the superior entities on which we depend: the company and/or the nation. It is our responsibility to be competitive and not being so could become a shameful trait of our personality. Once millions of men sacrificed their lives to save their homeland. Today billions of men sacrifice theirs for the competitiveness of the system in which they live. They are invited or sometimes very brutally forced and they are formatted very early, in business or management schools, to become the wonderful competitor that society needs. The cult of competition and competitiveness, the admiration of showbiz stars or that of great business leaders, the fascination for luxury and power constitute a norm from which it is quite difficult to escape. It is quite striking that in the film "The Wolf of Wall Street", the character played by L. Di Caprio, the archetype of the selfish, unscrupulous man, obsessed with increasing his wealth and power, still ends up generating a certain form of sympathy, at least a form of fascination and not disgust. This says a lot about the norm we have internalised. The so popular concepts of "winner" and "loser" are another expression of it. There is little doubt that they delineate in the collective unconscious what is valued and what is not. Unfortunately, I fear that, for many, it is better to be an unscrupulous and selfish winner than an empathetic loser, concerned with social justice and ready to sacrifice his comfort of life and material ease to preserve the environment.

The financial capitalisation company "Enron" invented the term "killer", a positive term associated with their most competitive employees, ready to renounce the most fundamental ethical values to enhance the company. It is symptomatic that a term as negative as "killer" has now become a quasi-praiseworthy term. The management of CompUSA encouraged their staff to see their colleagues as enemies,[15] implying that competitiveness is preferable to cooperation. Goldman Sachs supported the policy of "hard knocks and sharp elbows".[16] At the level of national policies, many marvel at the competitiveness of Germany ignoring, voluntarily or not, a

[15] Puffer (1999).

[16] Translation of "hard-knuckled and sharp-elbowed".

wage policy that has produced millions of poor workers. Our society has instituted competition as an ideal and those who do not adhere to it are the losers, the weak, the "losers": the dregs of a humanity that has not yet adapted to the violent charm of adrenaline and chronic insecurity necessary for the proper functioning of the system, it is thought in the most cynical managerial doxa.

I would like to show here that if our propensity for competition is a fundamental psychological property with an undeniable biological basis, it is nevertheless only one characteristic among others in human psychology, a characteristic that the evolution of our societies over the past few thousand years has established as a norm, a norm that is actually contingent and very unevenly present in each of us. You have probably noticed that people in your surroundings differ on this component of human nature. Some rejoice at the idea of confronting others and all situations in everyday life are experienced as possible spaces for competition, while others are totally refractory to it. Research conducted in political psychology and social psychology[17] has revealed the existence of stable personality traits associated with individuals seeking or rejecting competition and valuing or rejecting social hierarchy. From statistical and correlational studies, the authors have thus identified a psychometric dimension on which each subject can be evaluated. This dimension is called "social dominance orientation" or SDO.[18] SDO is defined as the degree to which individuals support social hierarchy, the domination of "inferior" groups by "superior" groups. Everyone can thus be located on a so-called SDO scale based on the answers they give to a large number of items—questions about their political orientation and their position on socially important issues—which allow individuals to be well discriminated on this single dimension, that is, independently of other traits of their personality. Here are some examples of questions, among a very large number, constituting this scale: "It's probably a good thing that some groups are on top and others are on the bottom"; "Some groups of people are simply inferior to other groups"; "We should not try to ensure that every group has the same quality of life"; "To get what you want, it is sometimes necessary to use force against other groups"; "winning takes precedence over the way of winning...". There is a continuum with, at the extremes, SDO+ subjects, who consider that competition is inherent and even necessary to life, and SDO-subjects, who avoid competition and favour cooperation.[19] SDO+ subjects are characterised by a set of other associated personality traits and by well-identified political choices and attitudes that clearly distinguish them from SDO- subjects. They are generally not very empathetic, not very altruistic and not very sensitive to the difficulties that others encounter. Individualism is strongly assumed. Machiavellianism can be justified if it allows one to achieve one's objectives, just as moral values should not be an obstacle to their realisation. Social hierarchy would be legitimate insofar as some humans are more powerful and intelligent than others. Therefore, the domination of social groups over others would be

[17] Feygina (2013).

[18] Translation of social dominant orientation. This dimension was first highlighted by Pratto (1999).

[19] Duckitt (2001).

natural, as would inequality.[20] The goal of increasing equality between individuals would be a counterproductive utopia because inequality is a natural fact. Meritocracy is often alleged to justify extreme levels of inequality. On the political chessboard, SDO+ subjects are conservative and are located on the right of the political chessboard. In the USA, surveys have shown that the level of SDO is much higher among Republicans than among Democrats. SDO+ subjects are more sexist than SDO-. They are very clearly nationalist, more frequently xenophobic and are more inclined to use violence when it comes to preserving their own interest and the advantages of their group of belonging or their nation. Racism, as an archaic expression of a preference for one group over another group—national preference is an example—is a notable trait of SDO+ personalities.[21] This is consistent with the fact that racism is based on the belief in the natural superiority of one group of humans over another group of humans.

It should be noted here the very strong correlation between attraction for competition and acceptance of inequality, confirming that these are indeed the two sides of the same psychological and therefore political reality. Moreover, and quite crucially, SDO represents a stable personality trait which is a key element of the so-called social dominance theory (TDS).[22] According to this theory, all complex societies are organised into dominant and dominated social groups. Dominant groups hold disproportionate power and use it against dominated groups to preserve the advantages linked to their status. Social hierarchy has a real adaptive interest and is therefore observed in a large number of social animals. Hierarchically organised societies have a real adaptive advantage because the acceptance of hierarchy reduces interindividual conflicts. Moreover, in case of conflicts with rival groups, hierarchically organised groups are more efficient, which is why hierarchy plays such a large role in the military world. Among humans, until the Neolithic period, the only social hierarchy that dominated was that concerning sex and age. Regardless of the type of society studied since the Neolithic period by anthropologists, adult men have always monopolised power. There is no known matriarchal society where women would have held power—military, financial, political—equivalent to that of men.[23] But with the Neolithic period and the ability to produce more than necessary for the survival of the group and therefore to appropriate a variable part of this production, an arbitrary social hierarchy appeared in the sense that it was no longer based on the "natural" distinction of sex or age. This new hierarchy, making inequality a structuring principle of these new societies, often imposed itself with extreme violence. Slavery, feudalism, the subjugation of one ethnic group by another dominant ethnic group during colonisation are classic examples. It is undeniable that with the emergence of inequality in a world now more governed by competition than by solidarity and cooperation, it is the SDO+ type personalities who have best played their cards.

[20] Pratto (1999).

[21] Altemayer (1998).

[22] Sidanius and Pratto (1999).

[23] Busch (1990); Sidanius and Pratto, (1999).

What is remarkable in the TDS is the fact that it is accepted by both the dominant and the dominated as if there was something inevitable about one group dominating another. This belief is consciously or unconsciously supported by founding myths, stereotypes, implicit beliefs that are there to legitimise social hierarchy and inequality. In the case of racism, the myth consists in admitting a "natural superiority of a race". In the case of Indian castes, superiority is relayed by religion, just as in Christianity, the legitimacy of royal power was associated with the supposed direct link between God and the king. In our contemporary societies, the myth of meritocracy is there to justify immense inequalities even though an individual's wealth is almost exclusively determined by that of his parents. Another myth is that inspired by social Darwinism, the concept that the struggle between individuals and the resulting inequality are only the social form of natural evolutionary mechanisms leading to the evolution of species: to the survival of the strongest and the elimination of the weakest. What is particularly interesting in the TDS is that it reveals, in a completely empirical way, that the hierarchical and unequal organisation of society is of course actively supported by the dominant class, but, paradoxically very often also, by the dominated class. Both accept in reality another myth, the so-called "just world myth".[24] According to this myth, a contemporary form of Voltaire's best of all possible worlds, if the world is as it is, it is because it must be as it is. Existing conditions, even when they are disadvantageous for the dominated group, are accepted and legitimised by the mere fact that they exist. There is a strong propensity of humans to convince themselves of the legitimacy of the situation in which they live. This process of rationalisation actually provides a major psychological satisfaction. It has the advantage of reducing the pain and psychological suffering that could emerge from the realistic perception of an unjust system of which the dominated are in fact the consenting victims.[25] It is better to convince oneself that things are not so bad, especially when the power of the dominant does not seem to be able to be contested. In this way, members of the dominated group participate in their own domination and the maintenance of the status quo.

This phenomenon gives rise to very surprising forms of acceptance of injustice. For example, during apartheid, many black maids found their situation legitimate and considered themselves lucky to be able to benefit from the help of their master.[26] Even worse, among the prisoners of Nazi concentration camps, many had to adjust their representation of the horror they were living in order to psychologically survive the absurdity of their situation. Some managed to legitimise the cruelty of their torturers and justify the obedience to which they were forced.[27] In the unequal system that is ours currently, the dominant as well as many of the dominated, admit stereotypes of the type "the poor are poor because they are less intelligent, because they are lazy and the rich are rich because they are enterprising, hardworking and

[24] This is the theory of belief in a just world initiated by Lerner (1980).

[25] Barclay and Saldanha (2015); Lerner (1980).

[26] Durrheim et al. (2014).

[27] Elkins (1967).

intelligent". The stereotype is so powerfully internalised that it leads to a freely consented submission blocking any possibility of rebellion.

The TDS that predominates in our society is characterised by a number of beliefs that the dominant have managed to instil in our representations and that the dominated have ended up internalising.[28] Among these are: the necessity of economic growth, faith in the effectiveness of science to solve all our environmental problems, the importance of preserving the current social order, the positive value of individualism and competition, the need to limit government controls, faith in an abundance that can only increase, the belief that the unregulated market is optimal in relation to all the problems we can encounter today... All these representations should lead to the maintenance of the current system which would not only be the best, but the only possible one. The Belief in the necessity and possibility of infinite growth is an integral part of the SDO and individuals who adhere most to the SDO are also those who show a very low sensitivity to the environmental crisis.[29] They frequently hold anti-ecological positions and tend to downplay the importance of the impending crisis. Among them, we find the most climate skeptics, a climate skepticism that is a classic form of psychological denial in the sense that their ideology unconsciously constrains their cognitive activity in the field to the point where they can no longer correctly process the information they have.[30] They believe that the environmental crisis can very well be overcome without changing anything about the current social and political system and that, in the event that a choice must be made between economic efficiency and environmental protection, it is economic efficiency that must prevail. It is also among them that the desire to dominate nature and a strongly anthropocentric conception such that the lives of other living beings are of little importance compared to the importance of the infinite pursuit of human development is most strongly expressed. Their desire to maintain the status quo, a status quo that is often favourable to them, generally leads them to deny the existence of a real ecological problem.[31] From a certain point of view, their rational analysis gives way to a form of magical thinking that strangely convinces them that infinite growth is compatible with the finiteness of the planet.

It is very clear here that there is not only a strong causal relationship between competition, inequality and environmental destruction, but this relationship is compounded by a psychological reality that is widely confirmed empirically: the most competitive subjects, those who support the maintenance of a strong social hierarchy are not only those who believe that inequality is a natural given, but also those who are least inclined to adopt environmental protection measures. The triptych of competition, inequality and environmental destruction is doubly confirmed. There is a fundamental incompatibility between liberal and capitalist ideology and the

[28] Dunlap and Van Liere (1984).

[29] Feygina et al. (2010); Feygina (2013).

[30] Hennes et al. (2016).

[31] Feygina (2013).

preservation of the environment and this incompatibility is empirically and psychologically confirmed: individuals who defend these values are much less sensitive to the ecological cause than others.

9.6 A Confirmatory Simulation

Several studies have empirically explored the actual behaviour and situation of SDO+ and SDO- personalities.[32] One of the pioneering studies in the field was conducted by social psychology researcher Bob Altemeyer in 1998. The experimental technique used first involved identifying the psychological profile of the subjects participating in these experiments. To do this, they answered a series of items and their responses allowed for a rigorous characterisation of their psychological profile. SDO+ individuals are leaders, ready to do anything to occupy the dominant functions that will allow them to satisfy their own interests or the interest of their group of belonging. Another psychological profile was introduced in these experiments. This is a profile that Bob Altemeyer called "Right Wing Authoritarianism (RWA)" and which I would translate as "authoritarian profile" because even if this type of personality often conforms to conservative right-wing values, it can perfectly conform to so-called left-wing values as was the case in some socialist or communist dictatorships in the twentieth century. To identify this profile, Altemeyer developed a questionnaire that allows it to be rigorously circumscribed. Here are two examples of questions from this questionnaire: "Our country desperately needs a powerful leader who will do what needs to be done to destroy the radical new ways and sin that are ruining us" or "The only chance our country has to overcome the crisis is to return to our traditional values, put strong leaders in power and silence troublemakers spreading bad ideas".[33] Three types of attitudes characterise these individuals, who will be called RWA+: their conservatism (obedient support for social norms and political authority), their authoritarianism (a firm attitude towards those who violate rules and laws and an obsession with order), their traditionalism (their unquestioning adherence to the social and religious values of the dominant group). Through correlational analyses, the authors have shown that this profile is frequently associated with a number of attitudes and ideological choices: ethnocentrism, racism, sexism, homophobia, dogmatism, refusal of ideological differences, pursuit of societal uniformity… SDO+ and RWA+ subjects are easily distinguishable. Among the most important differences, two of them are essential. RWA+ subjects need authority but do not necessarily seek positions of power. What matters to them is that there is a strong authority preserving the established order, not that they themselves assume this authority. They are therefore often docile followers ready for the most monstrous acts that their leader demands of them and abandon their analytical capacity when an order is given to them. The second difference is that RWA+ subjects have values, often traditional, whereas SDO+ subjects do not really have an

[32] Altemeyer (1998, 2003)

[33] Our translation.

ideology. Their pursuit of power and domination imposes an opportunism that is ill-suited to any ideological constraint. They are therefore not real conservatives except when conservatism allows them to preserve their interest and power.

Once the profile assessment is done, the experiment consists of studying how the subjects will behave in a game called the "Global Change Game". This game is a very sophisticated simulation of the current global situation and the problems we face: economy, military conflicts, biodiversity protection, global warming, natural resource management, world hunger, inequality management, health, education, justice… It was created by computer scientists with the help of experts in all the fundamental areas involved in the management of states and the planet. It is a very realistic game that simulates the current planetary situation as accurately as possible. Players are randomly assigned to one of the world's regions and behave as they wish with the aim of allowing the inhabitants of their region to live well. The results are quite enlightening. Firstly, when SDO+ or RWA+ subjects are staged, the result is disastrous: in both cases, these individuals have reached a nuclear holocaust that ended almost all of humanity. The natural environment was completely destroyed even before the nuclear war. The authors gave the same individuals a second chance to see if they were able to learn some lessons from the dramatic outcome they had reached. This time, they managed to avoid nuclear war. But more traditional military conflicts were just as devastating and problems related to inequality, the environment, malnutrition… were largely ignored. When the authors conducted the same simulation with SDO- or RWA- subjects, they observed that broad cooperation helped reduce the food problem. The environmental issue also received particular attention and the number of deaths related to international conflicts or poor management of natural resources was greatly reduced.

Other experiments of this type have jointly staged SDO+ and SDO- subjects with RWA+ and RWA- subjects who were brought to team up. If the association of RWA- and SDO- subjects proved optimal, the association of RWA+ and SDO+ profiles proved catastrophic. In fact, when this type of subject is associated, most of the time it is the SDO+ subjects who take the leadership over the RWA+ subjects, which is consistent with their taste for competition and their obsession with power. Very quickly, the RWA+ subjects scrupulously submit to the directives of the SDO+ subjects, even when these directives lead them to make unethical decisions harmful to humans and the environment. When such subjects are confronted with the Global Game, the exploitation of workers from poorer countries and the degradation of the environment of these countries is frequently observed, especially when they lead to the assertion of the economic superiority of the country they lead. The ethical dimension completely eclipses the objective of economic success of its group. As Altemeyer noted, the association of RWA+ and SDO+ profiles is lethal for humanity. The couple formed by President Trump and his Vice-President Pence seems a perfect illustration of this deadly dyad. Trump is the caricature of the SDO+ profile. He is first and foremost an opportunist who cares nothing for ethics and is guided by no ideology except that which allows him to conquer and retain power. Trump was indeed a Democrat before he was a Republican and does not hesitate to lie when it can serve him. The antics he has known throughout his life illustrate the little value he places on any form of ethics. Conversely, Pence, clearly RWA+, is clearly at the service of his master. He presents himself as being, in order: Christian, conservative and Republican. His essential values are order, security and tradition. Their association at the head of the American state is probably extremely dangerous. In relation to the environmental crisis, it is probably the worst type of governance one can imagine.

Another very interesting result concerns the behaviour of subjects who are both SDO+ and RWA+: subjects who combine the mental characteristics of these two profiles. Subjects who are both SDO+ and RWA+ are extremely dominant, ambitious and insensitive to the social difficulties of others or to the environmental issue. They combine the worst elements of these two profiles and represent a real danger according to Altemeyer.[34] Their thirst for success and domination is coupled with violent authoritarianism. They correspond to the very type of authoritarian leader and intolerant towards his opponents and rival states. In the context of the global change game, the outcome was inevitably that of the rapid destruction of the planet and humanity.

The last result, probably the most dramatic of these simulations, is that SDO+ individuals are so power-hungry that they almost systematically impose themselves on others. Worse still, when SDO- and RWA- type individuals manage to function rationally and cooperate for the common good and for the greatest number, it is enough to integrate an SDO+ individual into the group for this individual to manage to seize power and very quickly degrade the functioning of this virtual micro-society. As Altemeyer notes about SDO+ and RWA+ individuals, although they are relatively few in number, they probably have a considerable impact in our societies because of their ability to impose their domination. and to have a battalion of RWA+ type individuals who will submit docilely to their power. The history of humanity is undoubtedly largely imposed by these subjects.

9.7 The Real World and the Simulated World

The results of these simulations, staging different psychological profiles in the virtual management of the planet, reveal a number of essential points. First, they perfectly illustrate the close relationship between competition, inequality and environmental destruction. Even if it is only statistics, it is clear that the most competitive individuals and power-hungry, those who revel in the current economic system, are also those who admit and even wish for inequality and those who attach the least importance to the ecological dimension. The world they generate is a conflictual world, deeply unequal and ecologically degraded. It is generally the world that we are experiencing, thankfully short of nuclear holocaust. Secondly, if we take these simulations seriously then the situation in which we find ourselves today, both

[34] Altemeyer (2004).

politically and humanly as well as environmentally, is contingent. It is contingent because it is likely that if the world was governed by subjects neither SDO+ nor RWA+, the state of the planet would be very different. As psychologists and ethologists have shown, humans are neither the violent, selfish, opportunistic and unempathetic beast nor the rational, altruistic and respectful angel that some imagine. We are made of all this at the same time as these psychological components are present in each of us and in each of the societies we have built. The whole problem seems to be that one or other of these components can take over the other depending on multiple parameters that we do not control or little. Obviously, the fact that the SDO+ type individuals have naturally been led to occupy the positions of power throughout our history has certainly contributed to giving our humanity the characteristics that we deplore today. And if we judge by the profile of many current government leaders, often democratically elected, including at the head of the most powerful states, the global situation could well be very critical. The positive point, on the other hand, is that the rather catastrophic social and environmental situation in which we find ourselves is not necessary.

The only methodological reservation that can be made about the results of these simulations, and the conclusions we have drawn from them, lies in the fact that a simulation, even very realistic, is far from being able to represent the complexity of reality. The fact that, in these simulations, nuclear war occurred quite quickly when the world was managed by SDO+ individuals clearly indicates that implication and behaviours are not constrained in the same way in the virtual world and in the real world. Dying virtually from a nuclear war is nothing like dying really. However, I do not think that the overall result is contestable. The management of the world by individuals wishing to perpetuate an economically conflictual and unequal system inevitably leads to the psychological distress of the dominated and to the destruction of our environment.

Very similar results were obtained in the context of Hardin's tragedy of the commons. The experimental approach consisted of asking individuals constituting a sort of small human community to organise themselves to preserve a common good—the atmosphere for example—each individual being naturally subject to two contradictory pressures: that of preserving the common good and that of defending his own and only interest. In other words, it was a question of putting in tension the altruistic and selfish component. Analogously to previous experiments, it was a question of evaluating the type of treatment of the situation posed according to the level of altruism, empathy and narcissism of the protagonists. The evaluation of these psychological profiles is carried out using rigorously constructed standardised tests. The narcissistic component[35] was introduced because we know that such individuals have, more than others, need to display their success. It is likely that this leads them to selfish behaviour incompatible with the coordinated management of the common good. As in the previous simulations, the impact of the individual

[35] Campbell et al. (2005).

psychological profile is absolutely decisive.[36] When communities of selfish, unempathetic and narcissistic individuals are constituted, the management of the common good ends exactly as Hardin had supposed. Selfishness taking precedence over altruism, the exhaustion of the common good inevitably occurs, placing all individuals in a rapidly catastrophic position. When it comes to altruistic, empathetic and little narcissistic individuals, cooperation agreements are quickly found and the common good is systematically preserved. A very interesting result, which replicates that observed in the experiments manipulating the level of SDO and RWA, when altruistic individuals are associated with individuals prioritising their personal interest, the former tend to align with the behaviour of the latter, so that the presence of a single selfish individual can cause the collaboration to fail if he is not excluded from the group.

All these empirical researches reveal the existence of a significant interindividual variability. Human nature is not one or, at least, can give rise to radically opposed behaviours. Moreover, the individuals most adapted to competition, the most ambitious, the most motivated by personal success are also those who will achieve the highest social status and occupy the most influential hierarchical positions, those which precisely determine a social norm to which the other individuals will eventually conform. This is what we call social desirability: we tend to value the social norm of the moment, usually supported by the dominant group, whatever it may sometimes be, and even when it does not correspond to our own values. The extreme sociability of humans leads them to appropriate norms that the group implicitly imposes without us being aware of conforming to this norm and without even being aware of its contingent nature. To take an extreme example, under the Third Reich in Germany, but also in France, anti-Semitism was a terrifying norm that many appropriated in good faith, participating freely and with great zeal in what was a terrible tragedy.[37] Today, neoliberal ideology, the cult of competition and competitiveness, admiration for showbiz stars or for great business leaders, fascination for luxury and power constitute a norm from which it is quite difficult to escape. It could therefore well be that the most salient cardinal values of our culture are not those that would allow us to properly manage the common good that is our planet. The unchallenged domination of the liberal economy since the collapse of the socialist alternative and the fact that it has now conquered the imagination of citizens all over the planet constitute a very unfavourable cultural context, especially as its supporters, skilfully or naively, try to persuade us that the solutions to the environmental crisis will come only from technological progress. This cultural context plays in a direction opposite to that which prevailed in the Upper Palaeolithic when the selfish or excessively opportunistic individual was quickly ostracised. Today, it is the individual who refuses competition, submission to the ideal of material wealth and who rejects the selfish norm who risks being marginalised. Without a radical reversal of this norm, it is unlikely that we will be able to solve our environmental and social problems.

---

[36] Batson & Moran (1999); Sheldon and McGregor (2000).

[37] See the concept of the banality of evil by Arendt (1959).

References

Altemeyer, B. (1998). The other "authoritarian personality". In M. Zanna (Ed.), *Advances in experimental social psychology* (Vol. 30, pp. 47–92). Academic Press.

Altemeyer, B. (2003). What happens when authoritarians inherit the earth: A simulation. *Analyses of Social Issues and Public Policy, 3*, 161–169.

Altemeyer, B. (2004). Highly dominating, highly authoritarian personalities. *The Journal of Social Psychology, 144*(4), 421–447.

Arendt, H. (1959). *The human condition*. University of Chicago Press.

Barclay, L. J., & Saldanha, M. F. (2015). Recovering from organisational injustice: New directions in theory and research. In M. Ambrose & R. Cropanzano (Eds.), *The Oxford handbook of justice in work organisations* (pp. 497–522). Oxford University Press.

Batson, C. D., & Moran, T. (1999). Empathy-induced altruism in a prisoner's dilemma. *European Journal of Social Psychology, 7*, 909–924.

Bjørn, A., Kalbar, P., Elsborg Nygaard, S., Kabins, S., Jensen, C. L., Birkved, M., Schmidt, J., & Hauschild, M. Z. (2018). Pursuing necessary reductions in embedded GHG emissions of developed nations: Will efficiency improvements and changes in consumption get us there? *Global Environmental Change, 52*, 314–324.

Bourguignon, F. (2012). *The globalisation of inequality*. Threshold.

Busch, R. C. (1990). *Family systems: Comparative study of the family*. P. Lang.

Campbell, W. K., Bush, C. P., Brunell, A. B., & Shelton, J. (2005). Understanding the social costs of narcissism: The case of the tragedy of the commons. *Personality and Social Psychology Bulletin, 31*(10), 1358–1368.

Duckitt, J. (2001). A dual-process cognitive-motivational theory of ideology and prejudice. *Advances in Experimental Social Psychology, 33*, 41–113.

Dunlap, R. E., & Van Liere, K. D. (1984). Commitment to the dominant social paradigm and concern for environmental quality. *Social Science Quarterly, 65*, 1013–1028.

Durrheim, K., Jacobs, N., & Dixon, J. (2014). Explaining the paradoxical effects of intergroup contact: Paternalistic relations and system justification in domestic labour in South Africa. *International Journal of Intercultural Relations, 41*, 150–164.

Elkins, S. (1967). Slavery and personality. In R. S. Lazarus & E. M. Opton Jr. (Eds.), *Personality* (pp. 395–420). Penguin.

Feygina, I. (2013). Social justice and the human–environment relationship: Common systemic, ideological, and psychological roots and processes. *Social Justice Research, 26*, 363–381.

Feygina, I., Jost, J. T., & Goldsmith, R. E. (2010). System justification, the denial of global warming, and the possibility of "system-sanctioned change". *Personality and Social Psychology Bulletin, 36*, 326–338.

Hennes, E. P., Ruisch, B. C., Feygina, I., Monteiro, C. A., & Jost, J. T. (2016). Motivated recall in the service of the economic system: The case of anthropogenic climate change. *Journal of Experimental Psychology. General, 145*, 755–771.

Kallis, G., et al. (2018). On degrowth. *Annual Review of Environment and Resources, 43*, 4.1–4.26.

Kubiszewski, I., Costanza, R., Franco, C., Lawn, P., Talberth, J., Jackson, T., & CAylmer, C. (2013). Beyond GDP: Measuring and achieving global genuine progress. *Ecological Economics, 93*, 57–68.

Lerner, M. J. (1980). *The belief in a just world: A fundamental delusion*. Plenum Press.

Milanovic, B., Lindert, P., & Williamson, J. (2007). *Measuring ancient inequality*. World Bank Policy Research Working Paper 4412.

Murtaugh, P. A., & Schlax, M. G. (2009). Reproduction and the carbon legacies of individuals. *Global Environmental Change, 19*, 14–20.

Pathou-Mathis, M. (1993). Paris, Odile Jacob.

Pickett, K., & Wilkinson, R. (2019). *To live happily, let's live equal!* The Bonds that Liberate.

Piketty, T. (2013). *Capital in the twenty-first century paperback*. Le Seuil.

Pinker, S. (2012). *The better angels of our nature. A history of violence and humanity*. Penguin.

Pratto, F. (1999). The puzzle of continuing group inequality: Piecing together psychological, social, and cultural forces in social dominance theory. In M. P. Zanna (Ed.), *Advances in experimental social psychology* (Vol. 31, pp. 191–263). Academic Press.

Puffer, S. (1999). CompUSA's James Halpin on technology, rewards, and commitment. *Academy of Management Executive, 13*, 29–37.

Pyysiainen, I., & Hauser, M. (2010). The origins of religion: Evolved adaptation or by-product? *Trends in Cognitive Sciences, 14*(3), 104–109.

Sheldon, K. M., & McGregor, H. A. (2000). Extrinsic value orientation and "the tragedy of the commons". *Journal of Personality, 68*, 383–411.

Sidanius, J., & Pratto, F. (1999). *Social dominance: An intergroup theory of social hierarchy and oppression*. Cambridge University Press.

Chapter 10
What Room for Manoeuvre?

Abstract Freeing ourselves from the determinisms that lead us to adopt behaviors incompatible with the preservation of the planet is extremely difficult. Obviously, it is impossible to change our internal determinisms. On the other hand, it is possible to influence external determinisms, i.e., those that come from our society. The proposal I make is to regulate individual consumption more or less radically and thus put an end to a fundamental principle that all societies have integrated: everyone has the right to consume as much as their financial means allow. I analyze the consequences of challenging this principle, examining the positive and negative outcomes that result. Challenging the dogma of totally free individual consumption implies a major anthropological and political revolution, the only one capable of preventing a massive collapse of our civilization.

We live, for some, in optimal economic and environmental conditions: I am undeniably one of them. Several billion other humans, including a large number of citizens of the richest countries, suffer from a difficult, humiliating, inhuman life and survive in a degraded and unhealthy environment. At the same time, to raise the standard of living of the poorest and to further increase that of the richest, we produce, consume and always destroy our environment more. We must now admit that this destruction, if it continues, will end up degrading the situation of everyone, including the current privileged. Some of the humans on the planet are aware of the tragedy unfolding before their eyes. But this very recent awareness has little impact on our individual and collective behaviours. The causes of this paralysis are numerous. They are primarily psychological and not technical as we would like to believe: if the problem were only technical, there is little doubt that human genius would be able to solve it. It is psychological or even anthropological because our psyche results from an evolution that has endowed it with extraordinary cognitive abilities giving it the power to act on its environment, and potentially to destroy it: the

T. Ripoll, *Human Psychology and the Destruction of the Planet*, Anthropocene – Humanities and Social Sciences, https://doi.org/10.1007/978-3-032-02080-2_10

Anthropocene is an obvious consequence. But a large part of our psyche, both cognitively and psychoaffectively and emotionally, remains globally that of a particular kind of primate, capable of altruism, complex social relations, empathy, but also of extreme aggression, an infinite need for domination over the world and its own kind, a psyche subject to multiple addictions including the one leading us to always covet what we do not yet have. To curb our violent passions and our huge potential for aggression, men first invented the Gods of the great religions, then the institutions. These have gradually given birth to democracies, but democracies that have integrated and even supported social and ethnic inequality. The societies that result from this today form a system, a system so constraining that it leaves relatively little room for manoeuvre for individuals as well as governments to take the necessary step aside to lead us towards a more just and ecologically viable society.

Until the nineteenth century, institutions, still perfectly unaware of the future environmental crisis looming, partly curbed the violence inherent in the human psyche. Only in part however, because they have, to varying degrees, always served the interests of the richest states and individuals and have never managed to put an end to incessant military conflicts. What is fascinating is that this society, despite being a vector of absurd and incredible inequalities, will not collapse under the onslaught of the left-behind or popular revolutions whose objective would be to put an end to great social inequality. It will collapse because of its intrinsic inability to solve the environmental crisis. The gravedigger of the current politico-economic system will not be the revolutionary fighting for the defence of his rights. It is the limits of the planet that will do the work and impose a revolution even deeper than all those that humanity has known so far.

The crucial question is how this revolution can come about. Although I am very suspicious of any revolutionary utopia, especially if it tries to impose itself by violence, it is indeed a revolution and not an evolution. It is a revolution on the psychological, economic and political dimensions. On the psychological dimension, since reducing our growth, the only viable solution to the environmental crisis, requires countering the most fundamental forces of the human psyche: those that have been at the heart of this work, those that deliver us to our terrible addiction to consumption and the pursuit of immediate pleasure. On the economic dimension since our economic system can only survive if it grows and growth is incompatible with the preservation of the environment. And as it is not feasible to reduce our growth without reducing the competition to which individuals, companies and states are engaged; without also reducing the level of inequality between the citizens of this world, it is the global political system that needs to be revolutionised. In short, this means not only deviating from a trajectory that we have been following since the Neolithic, but it also means, and perhaps above all, finding a societal organisation that would succeed in countering the psychic forces that have naturally led us to where we are now.

10.1 A Revolution as Easy as It Is Unlikely in the Short Term

From a certain point of view, such a revolution is within reach of hand. It would be enough if we individually agreed to change our behaviour and reduce our standard of living and consumption for the environmental crisis to be overcome immediately. The problem is that we are nearly 8 billion humans subjected, on the one hand to powerful psychological determinisms, on the other hand to a constraining politico-economic system, leaving no real possibility for individuals to modify their behaviour to the point of overcoming the environmental crisis. If part of the solution certainly lies in our ability to individually adopt ecologically-compatible behaviours, the overall solution to the problem cannot be individual. It can only come from institutions, states or aggregations of states. But what is true at the individual level is also true at the state level. Which state would be mad enough to reduce its growth if it was not assured that the other states with which it is in competition would do the same? This would be the assurance of a national decline, of impoverishing citizens who would not tolerate the relative decrease in their standard of living compared to that of the citizens of other states who would have continued to enjoy the material benefits of growth. We are therefore in a situation of absurd deadlock, an absurdity all the more glaring because we actually have, given our productive capacities, the means to allow every human being to live decently, materially and culturally.

10.2 There Is No Inevitability to the Environmental Crisis

There is no inevitability for three main reasons. The first is that while humans are constrained by their biology and by their evolutionary history, they have, more than all other species, the ability to escape from these determinisms as ancient as powerful because the evolved monkeys that we are as much subject to biological determinisms as to cultural determinisms. We must count on the weight of the latter to counteract what is most harmful in the human psyche. One only has to measure the impact of our culture on our individual and collective behaviour to be convinced. Humans can evolve at high speed because they have a unique and exceptional cognitive flexibility. The second is that the human psyche is not only the plaything of archaisms that we have analysed in this work. Humans are also naturally empathetic, altruistic, sensitive to injustice and inequalities, sometimes rational and capable of controlling their aggressiveness. The third reason is so obvious that we sometimes tend to forget it. It is that in reality, we have the technical means to solve both the social crisis and the environmental crisis… for a little while longer. It would be enough for us to agree to grow within the limits of what is physically and biologically compatible with the limits of the planet and to organise a much fairer distribution of the wealth produced. Producing less and consuming less poses no problem. Technically, it's very simple. As for a fairer distribution of wealth, there is a psychological and political obstacle still insurmountable that we will have to overcome 1 day to get out of the impasse in which we find ourselves. I am betting here that the environmental crisis will impose this revolution on us because at the moment when our survival on the planet will no longer be taken for granted, men will know how to find the resources to put in place a society that will no longer be governed by the psychological archaisms that continue today to lead us to disaster. In this sense, this crisis is a new opportunity to invent another society.

10.3 What Do We Have to Gain by Revolutionising the Current System?

Humans have limited rationality. This is indisputable. When it comes to evaluating their own short or medium-term interests, they make a quick calculation of risks and benefits and they only agree to make alternative choices if the expected benefits are greater than the expected risks. They will therefore only accept the necessary revolution if they are convinced that the risk/benefit calculation is individually favourable to them.

In this calculation, a determining element is the perception of the severity of the environmental situation. Perception being, by definition, the product of a psychological process, it results in a frequent gap between reality and perception. This gap is immense in the present case and leads to an underestimation of the environmental risk due to cognitive limitations[1] of the human psyche, particularly the difficulty in perceiving abrupt upheavals in geological time, and yet very slow and almost imperceptible in the time of a human life. This underestimation is further reinforced by the discourse of a large number of political or economic leaders who are too attached to the maintenance of the current system to accept to identify and reveal clearly the danger it poses to humanity.

This underestimation of risk also stems from the fact that the vast majority of humans do not have the scientific data necessary for a fair assessment of the situation, nor the cognitive skills to process them correctly. We will only accept a major change in our behaviour as the consequences of the ecological disaster become more prominent, as is the case with any person suffering from addiction. We will accept it all the more easily as we have serious and credible information and are able to process it correctly. Hence the importance of available information and the overall level of education necessary for its processing. On this point, we are far from the mark and it is to be feared that we will have to go very far, perhaps too far, in the destruction of the planet before the risk/benefit balance leads us to accept the need to radically change the organisation of society.

But the psyche of human is such that he needs to have in his sights positive projections to tip the risk/benefit balance on the right side. It is clear that we have not yet discussed these. What would we gain by shaking up the current economic-political system? Advocates of degrowth or agrowth traditionally evoke the concepts of happy frugality as a substitute for the pleasure of consumption. They are obviously not wrong as it is clear that happiness is only very weakly related to our

[1] Gifford (2011); Sorqvist and Langeborg (2019).

level of consumption or our material comfort. I personally know many people who adhere to and theorise this type of rhetoric. That said, very few actually put into practice what they proclaim loudly. First of all, it must be admitted that this is typically a position of the rich and almost exclusively of Western rich. Several billion humans, especially the poorest, would sincerely wonder if it is absurd humour or a form of romantic delirium of the wealthy if they were advised to live more frugally. On the other hand, most of us consider ourselves frugal in our daily lives while they contribute largely to the destruction of the environment due to consumption they can no longer even identify as such.

Advocating a form of happy frugality, encouraging people to work less, earn less, consume less in the perspective of increasing their quality of life is really too theoretical to be convincing, especially in a society where the possibility of working is diminishing as our productivity increases. Above all, happy frugality constitutes a form of relationship to the world that is a thousand miles from the "natural" psychic functioning[2] and, in fact, that of most humans. The whole purpose of this book has been precisely to show this. In short, it is this happy frugality that we need, it is indeed the renunciation of ever-increasing consumption that we must achieve, but it is not by suggesting to men to become frugal that we will achieve it. For the vast majority, the gain they perceive in adopting a frugal behaviour is almost nil. The association of the terms "frugality" and "happiness" undoubtedly makes sense, but the necessary link between these two concepts is absolutely not obvious. Frugality has no causal relationship with happiness: it does not lead to happiness. It is rather the opposite that is true. It is because one is happy and can have immaterial satisfactions—social, intellectual, artistic, relational, affective—and can therefore escape the fleeting and superficial satisfactions provided by strictly material consumption. Frugality is indeed an objective, but an objective that can only appeal to those who have the cultural capital to understand it and those, often the same, who do not suffer from a personal material situation so difficult that the prospect of increased frugality would appear to them as a strict nonsense, or even an additional oppression. Only a tiny fraction of humanity can be sensitive to the argument of frugality.

Undoubtedly, emphasising the importance of social ties could be more effective. Not effective enough to completely free us from our addiction to consumption, but necessary for the awareness of the impossibility of preserving a strong social link in a society based on a goal of increasing consumption and putting competition and inequality at the heart of its system. For what is most striking in our society is indeed its destructive power in terms of real social ties and not its current ersatz. Yet, among the most powerful determinants of happiness are the social link and the quality of the relationships we have with the human community. On a strictly empirical level, studies in both sociology and anthropology,[3] are convergent: the average time spent on social exchange has steadily decreased throughout our history. Social iso-

[2] Let's remember that "natural" does not in any way mean "good, adaptive, legitimate…". Violence is natural and it is condemnable.

[3] Sahlins (1968); Van de Velde (2018).

lation within our societies, and the scourge of imposed loneliness, have become major themes to the point that they now occupy a place in the political agenda. In the UK, a Ministry of Loneliness has been created to try to stem the extreme psychological suffering of nine million Britons. France made it a major national cause in 2011 because if loneliness generates well-identified psychological disorders, it is also responsible for a very large number of somatic diseases. Loneliness and loss of social ties kill as much as tobacco or alcohol. What value can we give to a society where the elderly find no other solutions than suicide to end their isolation?[4] A society where the time spent interacting with one's pet is longer than the time of exchange with other humans, a society where the time spent in front of a screen is far more important than that spent interacting with others, or even, a society in which, to fill the emotional and relational misery, billions of humans are looking, through a compulsive consumption, for a morbid palliative leading them to obesity, over-indebtedness, or the total loss of meaning?[5] How did the social animal par excellence that we are, an animal that cannot truly survive without real and physical contact with the other, come to such a painful situation? As often, even an unbearable and absurd situation ends up, by the mere play of habit and normalisation, appearing normal to us. But this semblance of normality does not alleviate the psychological suffering that results from it because it is indeed a whole part of the fundamental human needs that are undermined by the society we have created.

Psychologists, anthropologists and sociologists have finely studied the causes of this weakening of the social bond. They are obviously numerous and complex. But they all have something to do with the idolatry of competition and the extreme level of inequality we have reached. The destruction of the social bond is neither an epiphenomenon nor a contingent characteristic of our society: it is inherent in the principles that govern it. Living in a society that makes competition the heart of its functioning can only contribute to deteriorating the social bond. The forced individualism, the obsession with success and performance, the importance of putting each other in competition in most of the registers of our social life are hardly compatible with the strengthening of the social bond. Moreover, the resulting inequality, material but also cultural, tends to separate individuals, to compartmentalise them and leads us to perceive "the other" as potentially dangerous because of his or her otherness. We end up no longer making society precisely because we no longer live in the same society: our neighbourhoods differ, our leisure activities differ, our schools differ, our interests differ, our aspirations differ, our culture differs. These distinctions are not a problem in themselves because plurality is first and foremost a wealth. The problem is that these distinctions are imposed by our class membership, by our standard of living, by our material wealth, by our cultural capital, all characteristics that are imposed on the majority and which end up establishing barriers incompatible with living together. Not to mention of course that the

[4] Campeon (2011).

[5] See the extreme phenomena of Kodokushi and Hikikomori in Japan.

preservation of strong social ties requires both temporal availability and spatial proximity, both of which are undermined by the demand for professional performance.

Our society therefore not only destroys our environment. It destroys what contributes most effectively to happiness and our psychic balance. Destruction of the environment and destruction of the social bond are the two sides of the same reality: the sad price to pay for our material comfort, our wealth, our addiction to consumption. In the risk/benefit calculation, emphasising the importance of the quality of the social bond is probably more effective than emphasising the benefits of frugality. However, it will be necessary to concretely show how another type of society ecologically compatible could lead to reweaving a social bond that is constantly deteriorating.

10.4 Utopia Is Not Where We Think It Is

We are victims of an incredible and fatal utopia. A utopia that our descendants will later analyse as a fascinating illustration of the capacity of humans to deny the most obvious reality when it is not pleasant to them. This utopia is the cornucopian utopia, believing that technical progress will allow us to continue to grow without endangering the planet. The term utopia is somewhat paradoxical here since this society exists here and now. In fact, it is only a utopia in the sense that it will not be able to exist tomorrow. Whatever the technological progress that will occur, if the fundamentals of this society do not change, either our civilisation will hit the limits of the planet in terms of resources and viability in a too degraded ecosystem, or it will be confronted with large-scale conflicts due to the scarcity of resources necessary for our survival. In any case, the decline could be much more brutal than we usually imagine.

Because individuals can neither escape the forces that society exerts on them nor escape the determinisms that lead them to grow indefinitely at the individual as well as the collective level; since the destruction of the planet threatens in the short term the life of human as well as that of many living beings and even though growth no longer contributes to human happiness, the only solution consists in constraining the level of individual consumption. This is the heart of the solution that I propose and that I would now like to analyse.

10.5 What Solution?

Constraining the level of consumption may appear as a radical and utopian solution. In a sense, it is indeed the case because it is obvious that humanity is not yet ready: one does not defy like that what is at the foundation of human nature. But is it more utopian to decline proposals compatible with our survival and radically breaking with the established model or to conform to a model that works today but which we know is without a future in the medium or long term? Sooner or later, I am convinced that the solution that I advocate will appear to us as the only rational solution, unless humans prefer to the end the utopia of infinite growth with its consequences much more dramatic: those inherent in the impossible survival of the species on a degraded Earth and the large-scale conflicts that will result from it. necessarily. This solution is only the logical consequence of the analysis I have made throughout this book. I do not claim that such a solution is operational today. Yet it is the only serious possibility we have to escape a planetary disaster of which we will be the first victims.

What I propose is conceptually quite simple. We are nearly 8 billion humans sharing the common good that is our planet. If the word "share" has any meaning, then each human should have an equivalent right in terms of resource exploitation. Why, for example, when resources in fossil fuel are rapidly dwindling, should I accept that individuals use for their own use 10, 100 or 1000 times more than me and thus contribute, much more than me, to global warming that threatens us all? Would I have more right than others to exploit and destroy this common good simply because I inherited from wealthy parents, because I have a long education, that I have significant income or simply because I was lucky enough to be born in a rich country? Certainly not and I challenge anyone to legitimise such inequality in our rights to exploit the common good and, ultimately, to destroy it. The only justification for such a situation can only be that of tradition or usage: this is indeed the most mediocre and questionable of legitimations. Until now, inequality found a semblance of legitimacy through the fallacious concept of meritocracy. Above all, that some are infinitely richer than others had as "only" disadvantage of being detrimental to the poorest: it was just a "human affair". Moreover, democracies have always accommodated a scourge considered unconsciously as "natural" and which, in a certain way, is an integral part of the social contract. A norm so well integrated now that even the poorest have internalised it and largely accepted it. The big change, both philosophical and anthropological, is that we know today that inequality is not just a "human affair". It is also a "planet affair" because the richest do not live only at the expense of the poorest; they especially allow themselves the exploitation of a common good necessary for the survival of all. Nothing can justify that some, because of their status, their power or their wealth, arrogate the right to destroy what is necessary for the life of others. If planetary resources were infinite, if its exploitation did not butt against its own limits, inequality could be conceptualised as it has been for 10,000 years: the inevitable consequence of the natural superiority of some over others, with no other impact than the life of some is materially harder than that of others. But this legitimisation no longer makes any sense today because when an individual consumes as much as hundreds of millions of others, he destroys a resource and an environment that these hundreds of millions of other humans need. He therefore directly endangers their life and that of future generations.

Imagine a small community of humans being deposited on an island whose water resources are limited, so that the survival of this community is only possible if individual consumption does not exceed 50 litres per day. Suppose one of these islanders, infinitely richer than the others, has the financial means to pay and consume 1000 litres per day to feed his swimming pool, his Jacuzzi and keep his lawn perfectly green even in times of drought. If nothing constrains him in his consumption, if it only depends on his income, it is clear that the survival of the community will be seriously compromised. Is there any sense in granting him this freedom? Certainly not. Yet this is what we accept today on a larger island.

10.6 From This It Follows That...

Today, unlike yesterday, we know quite precisely the limits of the planet that cannot be transgressed with impunity.[6] These limits are well identified in terms of the ceiling level of production of the main chemical pollutants, the amount of CO_2 compatible with a temperature stabilisation and exploitation of the planet's natural resources. Given the number of inhabitants on Earth and given its limits, given of course also our current level of technological mastery, it is possible to calculate a personal account, a sort of drawing right that we should not individually exceed. It should be immediately clarified that this account makes sense only if it concerns our private consumption and not that which is imposed on us by our social function: our work. The energy and chemistry used by a farmer to exploit the thousands of hectares allocated to cereal production would obviously be excluded from his individual account. It should be noted that such an account has already been imagined in terms of our emissions of CO_2. Jean-Claude Jancovici was the initiator and similar accounts have been developed in many other countries. But, CO_2 and global warming are only a tiny part of the problem: the tree must not hide the forest. This account, which I would call Individual Ecological Impact Account (IEIA), should concern, not only CO_2, but also all the fundamental biophysical parameters on which the global balance of our ecosystem depends. This account could constitute an objective indicator of our individual ecological impact and would define the limit to which everyone should conform with the aim of preserving a viable ecosystem in the long term. We accept that our consumption is limited by the funds we have on our bank account. For what reason could our consumption not be limited by taking into account the resources of a common good on which everyone depends?

The implementation of such an account poses many technical and theoretical problems that I will only superficially address here, the objective not being to specify its operationalisation. While it is clear that we are increasingly able to identify the biophysical limits not to be exceeded to preserve the ecosystem, it will be much more difficult to quantify our individual consumption. The technical difficulties inherent in this quantification can be easily resolved given the development of new technologies in the digital field. The great difficulty is rather of an ethical and political nature. Such quantification requires monitoring of our consumer activity which

---

[6] Steffen et al. (2018).

leads to an intrusion into private life by private or public structures. This is potentially dangerous and serious safeguards would need to be put in place. That said, as cash is soon to disappear, even today, the tracking of our expenses and the nature of these expenses are easily accessible. Another technical problem that is not self-evident is that of the drawing right granted to each one. It is obvious that this right can hardly be identical for all. A man is a bigger consumer than a woman simply because his metabolism is more important. A person living at 2000 meters altitude will consume much more energy to heat than a seaside dweller. A worker who lives in the suburbs, unable to afford housing in the city centre, will consume more energy to get to his workplace. The list of parameters to take into account to set the ceiling of this individual account is long and will have to be the subject of a democratically validated consensus.

10.7 The Political Implications

The most "soft" use that can be made of such an account could be strictly informative and therefore non-binding. Each of us could assess our individual impact by measuring what we consume beyond the ceiling compatible with the preservation of the planet. This would be very useful as it would allow us to become aware of the reality of our personal consumption and its environmental impact. Many would discover that they are big consumers while they imagine themselves living soberly. Many of those who have one or more second homes, who have pets, who heat unduly and air-condition at the slightest excess of heat, who drive in powerful vehicles, who spend their Christmas holidays in the Caribbean... would realise the sobriety cure necessary for our collective survival. This would already be extremely useful and interesting—as we all underestimate our level of consumption—but insufficient. For all the reasons I have developed above, it is almost impossible for us to individually respect the ceiling of our individual account.

The more realistic alternative of a binding individual account has major implications that would lead to the establishment of nothing less than a new civilisational era. But such an alternative will encounter major obstacles in its implementation, probably as long as the threat of a systemic collapse is not concrete enough to collectively convince us to admit the necessity of adopting this proposal. Without constraint on our individual consumptions, there will be micro-changes, but these will be local changes that will have nothing systemic and will not deflect the current dynamic.

The first of these obstacles, probably the most important, is that of the necessary collaboration between states, which clearly means the end of economic competition. It is, indeed, totally impossible for a state to unilaterally decide to implement the IEIA in a binding manner as this would result in a significant weakening of the national economy with all the inevitably negative consequences this can have at the geopolitical level. As long as the global economy remains constrained by competition and economic warfare, such measures will be suicidal as they would inevitably lead to a decrease in the national growth of countries adopting these measures and would trigger a major recession, probably a source of serious social conflicts. Moreover, which citizen of these hypothetical ecologically virtuous countries would accept to reduce his material consumption if the other citizens of the planet continued to exploit the common good without limitation? Obviously, nationalism, patriotism, the famous national preference, and therefore the very concept of nation, all concepts derived from this archaic component of human nature that is parochial altruism, are the main obstacles to the adoption of the implementation of the IEIA, this individual account can only make sense if all the citizens of the planet have an approximately equivalent ceiling.

The binding approach of an identical individual account for all has two formidable implications that a large number of states, especially the most powerful and rich, would not be ready to accept today. The total cumulative ceiling of the IEIA corresponds to the consumption available for the almost 8 billion humans on the planet once the part inherent to the functioning of states to ensure their regalian missions has been subtracted from the total human consumption ceiling. This is a huge difficulty as states differ in their national management, in their "lifestyle" and in their real needs. This means that a prerequisite to the evaluation of the IEIA ceiling involves a negotiation between states about their own expenses. When we know that part of these expenses is not devoted to the well-being of citizens but to a defence budget intended to ensure their security or hegemony, we measure how far it is from the cup to the lips. This is just a final confirmation that international competition is a major obstacle to solving the ecological crisis.

The second implication is just as important. Once the part of consumption necessary to meet the needs of states has been subtracted from the total human consumption ceiling, there will remain a total quantity of available resources which, distributed among the nearly 8 billion humans, will give rise to the ceiling of the individual IEIA. It is clear that this ceiling will not be perceived in the same way by the inhabitants of rich countries and those of poor countries. It is likely that as a relatively well-off Frenchman, I would be forced to reduce my own consumption while a well-off Senegalese would not have the personal budgetary means to consume to the point of reaching the ceiling of his IEIA. Similarly, within a country, wealthy citizens would no longer be able to consume as much as their personal means would allow them to do so, while others, the most disadvantaged, would not be able to reach this ceiling due to insufficient income. This would of course be the case in France. Such a general scheme would obviously meet the opposition of the richest states, and within a state, of the wealthiest citizens because, at the individual level, the benefits inherent in a high income and wealth would largely disappear due to the adoption of the IEIA. What good would the fortune that some have if they can no longer squander it as they wish? What is the point of getting rich, if the standard of living of the wealthy no longer stands out as clearly from the standard of living of middle-class individuals? Faced with the loss of such privileges, there is no doubt that resistance to the adoption of the IEIA would be fierce on the part of those, a minority in reality, who have the absurd privilege of being able to destroy the planet more than others. This is another illustration of the fact that inequality and ecological disaster are intimately linked because, in the end, the wealth surplus of the wealthiest is always transformed into excess consumption and therefore destruction of the planet. Whether we like it or not, we destroy the common good that is the Earth in proportion to the financial means we have. The adoption of the IEIA would thus put an end to one of the most extraordinary absurdities of our humanity: billions of individuals consume excessively without it having an impact on their level of happiness and this consumption is at the expense of the common good that is our planet and therefore at the expense of the life of all humans.

10.8 Economic Consequences

Clearly, such a upheaval in our freedom to consume would have major economic consequences. It would necessarily result in a global decrease in growth and an even more massive decrease within certain states, particularly those with the highest GDP per capita today. In short, it would lead to a recession: the most dangerous threat from the point of view of any orthodox economist. In reality, recession is indeed a situation that our economies and societies do not know how to deal with because recession means unemployment, global impoverishment and social conflicts. To take an example as concrete as it is revealing of the absurdity of our economic system, if we no longer allow billionaires to have terribly polluting yachts of pharaonic dimensions built, then shipyards will close and engineers, technicians, workers will lose their jobs. This scenario applying to all areas of activity, it will result in a decrease in the global production of wealth. But this classic scenario is only dangerous because of the incredible level of inequality we have reached and the organisation of work we have adopted. In times of recession, wealth production decreases and affects both the wealthy and the disadvantaged classes, but its impact is dramatically more important for the disadvantaged classes. The devastating effect of recession is not the same for someone who lived in great comfort before it occurred as for someone who was already barely surviving above the poverty line. In reality, our productive capacity could be sufficient to allow all of humanity to live in a quite decent, even comfortable way, even if the appreciation of "comfortable" must vary considerably with one's own standard of living. It is likely that Jeff Bezos or Bill Gates would find my house indecently austere and unworthy when I already consider it unnecessarily large and comfortable. The question is simple. Should we allow the wealthiest to consume freely and outrageously in view of the planet's limits[7] when billions of humans cannot even meet their basic needs, or should we set a consumption ceiling so that all humans could live in decent conditions without transgressing the planet's limits? Maintaining our growth rate at a level compatible with the preservation of the planet would have important and formidable economic consequences that could only be overcome once the production of wealth was better

[7] It is considered that the richest 1% of the planet consume as much as the poorest half of the planet.

distributed than it is today. The implementation of the IEIA would directly contribute to this because capping our personal consumption would provide states with the opportunity to redistribute to the most disadvantaged what the wealthiest could no longer consume. Therefore, a large part of the problems associated with any recession should be partly absorbed by the opportunity given to the poorest to increase their own level of consumption by virtue of the redistributive capacities of states. Beyond this particular point, it is the whole economic logic that would be metamorphosed and therefore the one that presides over our relationship to society and life. The sacrosanct universal objective of personal enrichment that structures our lives willingly or by force would no longer make any sense and this would have profound repercussions in the development of each individual. The simple fact of limiting our freedom to consume would completely reshuffle the cards by profoundly modifying the nature of the relationships we establish with others. The scourge of exacerbated competition and the scourge of inequality would be immediately diminished, which would lead to the disappearance of a good part of the ills we currently suffer from.

It is also clear that such moderation in global growth would have significant consequences in terms of employment because, as productivity necessarily continues to increase and consumption stabilises at ecologically viable levels, it is certain that a large number of humans would be deprived of work. All of this is the reason why the theme of reducing working hours, like that of universal income, holds an essential place among supporters of a global moderation of growth. I will simply remind you that, for a Paleolithic human, the time constrained by hunting, making objects, making clothes, building houses..., what can be broadly assimilated to working time,[8] was around 5 h per day. This time reached about 12 h for employees at the end of the eighteenth century and the beginning of the nineteenth century even though productivity was enormous compared to what it was in the Paleolithic. And yes, to consume more, men had to work much more, to exhaustion for many of them, including even children who became slaves. The very fact of limiting growth while human productivity will continue to increase necessarily imposes a limit on working time and sharing it. But as enviable as this may seem at first glance, it will not be without problems. First because humans fear idleness more than anything else, then because we have so integrated that work was constitutive of our life, our achievement and our personal fulfilment, that the prospect of a less laborious society scares more than it attracts in reality. And yet, the logic is unassailable. If we accept a reasonable level of global consumption and if productivity continues to increase then the reduction of working time is absolutely inevitable, as is the implementation of some form or another of universal income.

[8]The concept of working time is recent and these comparisons should be taken with caution as the concept of "work" had no meaning in the Paleolithic.

10.9 Concrete Consequences and Psychological Consequences

The consequences of the new relationship to the world, to others and to the planet, that would be caused by the adoption of the IEIA, are gigantic. It is precisely because of the enormity of this change, the only one compatible with our survival on Earth, that we find it so hard to admit. The rupture would be so great compared to what has structured our lives up to now—competition, inequality, total freedom to consume and the obsession with growth—that our resistance is total and in proportion to a shock of representation that our cognitive system struggles to integrate. In fact, such a change requires escaping from our animality, escaping from the archaic forces that lead us, like all species, to grow and try to occupy the largest part of their ecosystem. These consequences are therefore countless and I will limit myself here to analysing the main ones.

The first very concrete consequence is that a very large number of humans would be confronted with a decrease in their standard of living. It is quite easy technically to evaluate the amount of wealth produced by the totality of humanity but it is much more difficult to evaluate the average income or wealth of each inhabitant. Without going into the detail of these economic statistics,[9] it is necessary to recall some orders of magnitude. The total average wealth of an adult individual is of the order of 75,000 dollars[10] but varies considerably depending on the country in which one lives. Concretely, let's remember that a French person earning 2000 euros net per month is part of the 15% of the wealthiest humans globally. We can turn the figures as we wish, but the orders of magnitude speak for themselves. Given the global wealth produced, given the fact that this wealth should either stabilise or decrease, the standard of living of the inhabitants of rich countries should inevitably decrease. Our very high standard of living results both from a devastating overexploitation of the planet and from a perfectly unequal appropriation of its wealth. Ending this is an integral part of a genuine ecological revolution and it would be dishonest to make the wealthy of this world believe that this will not have an impact on their daily lives. It is likely that many of those who campaign for a drastic reduction in current inequalities and for taking into account the limits of the planet would revise their ecological commitment once they have become aware of what this implies in terms of modifying their own standard of living.

The other major consequence will concern our mode of consumption. Regardless of our income level, choices will have to be made. It is clear that a large part of ostentatious consumption will be abandoned even by those who are so happy to display their wealth. Here are two examples. Living in a 200 m^2 house when you live as a couple and having a second home with a swimming pool that you use at best 3 weeks a year, would require giving up multiple other positive aspects of life.

---

[9] Global wealth report 2020, research institute credit Suisse. https://blogs.worldbank.org/en/opendata/new-country-classifications-income-2019-2020-update

[10] The median wealth would be very different given the very unequal distribution of these riches.

It is not certain that this type of luxury will be maintained in view of what it will impose in terms of other renunciations. Deciding to drive a very consuming SUV and also having a city car, a motorcycle and a real off-road vehicle will quickly appear as an obstacle to other satisfactions probably more fundamental in terms of their impact on our quality of life. Generally speaking, a large part of our ordinary consumption will be abandoned because the fact even having to be selective in our consumption will lead us to rank them and perceive many of them as perfectly trivial. This will lead to asking a fundamental and very simple question that we would do well to ask ourselves each time, for example, when we acquire a new good: will this acquisition really contribute to my happiness? This will also lead us to systematically evaluate the relationship between the importance of the environmental cost, in terms of draw on the IEIA, and the gain in terms of quality of life. It is not certain, under these conditions, that we will continue to consume lobsters caught in the Caribbean or strawberries in winter. It is also not certain that we will want to acquire a new mobile phone when the previous one was working perfectly, just as it is not certain that we will keep it switched on 24/7 without real necessity. The list is long of the vital adjustments we would make as our current consumption is disconnected both from real satisfactions and from what it implies at the environmental level.

It is quite likely that the application of the IEIA will divert us from strictly material consumption and engage us more in forms of consumption maximising our social ties or of high cultural value: what we really need for our psychic balance and which effectively contributes to our well-being. It is clear that the current movement of collaborative economy would be considerably boosted by the adoption of the IEIA. To take just one example, the IEIA would probably no longer allow the individual acquisition of a yacht and this is all the more reasonable as a yacht is used on average 1 week per year, which is an exorbitant purchase and operating cost if we relate it to the number of sailing days, not to mention of course the monumental ecological impact of the boating industry. In a society constrained by the IEIA, individuals will be led to share their acquisitions, to manage them collectively: this would be humanly enriching and ecologically indispensable.

In the leisure economy, an economy that has experienced the greatest growth in recent years, we will witness a upheaval of practices. Again, I will only take one example; that of these fleeting trips by spectator consumers, capable of travelling tens of thousands of kilometres to see a spectacular site that they will hastily photograph without even having had time for an exchange with a local resident. Long-distance travel will become exceptional and will unfold over time, which will be possible as the time devoted to work will be considerably reduced. But the most likely is that citizens will engage in infinitely less spectacular leisure activities, and above all, less catastrophic on the ecological level. Regarding travel, the discovery of places, like individuals, is infinitely richer when it is done on foot or by more ecological and slower means of transport. This would result in an increase in autonomy and freedom of action compared to the standardised organised trip which dramatically constrains the pleasure of exploring and the possibility of impromptu encounters. Above all, cultural leisure will gradually replace consumption leisure.

Doing theatre, playing music in a group, singing in a choir, drawing, reading, painting, playing with others... are all very low-consuming and very enriching activities. There is no doubt that they would be widely practised not only because of their low ecological impact, but also because of the enrichment of social ties they allow. In short, the constraint linked to the implementation of the IEIA would probably result in reducing activities where the citizen is a passive, fast, often solitary and not very autonomous consumer and in increasing high cultural value activities, promoting exchange or sharing, activities requiring the individual to be more autonomous and more active. In reality, our current excess consumption partly compensates for the impoverishment of our social relations: it corresponds to a material substitute for the vital need we have for social interactions, a need extremely undermined in our society. By reducing the possibility of consuming in the traditional sense of the word, it is likely that the adoption of the IEIA will intensify and improve the quality of the social link. This shift from material consumption to socially rich activities is all the more likely as the IEIA implies a reduction in working time, a decrease in the level of inequality and therefore the attenuation of antagonistic relations induced by the climate of competitiveness that dominates today.

10.10 The Role of Science and Demography

In Western countries, we benefit from an exceptional standard of living that we will have to reduce if we want to preserve the planet. The whole question is to ensure that this reduction in standard of living is neither too abrupt nor too radical. For this, we have two levers by virtue of Ehrlich's law: I = P.A.T. For a fixed global ecological impact (I), the individual level of consumption and wealth (A) can only increase if the population (P) decreases and/or if technological efficiency (T) increases.

The first lever is therefore science and the resulting technological innovation. The latter has indeed continued to progress but, to date, only with the perspective of growing ever more. This dynamic of scientific innovation must be pursued, no longer with the objective of economic growth, but with that of improving our living conditions within the limits imposed by the necessity to preserve the planet.

The second lever, a taboo subject if ever there was one, is therefore that of demography which must be dealt with rationally and without prejudice. From a strictly quantitative point of view, it is clear that the ceiling of the IEIA could be higher if the number of humans on Earth was lower. Indeed, if we admit that, at constant technological efficiency, we cannot exceed a ceiling of production and consumption determined by the biophysical limits of the planet, the The ceiling of the IEIA will remain unchanged if the world population does not change, will increase if it decreases and will decrease if it increases.[11] This is the simple equation we face, which is why the issue of demographics remains essential.

This issue has quite divisively, even caricaturally, pitted supporters of a total absence of population control against those who see in unregulated population growth the primary cause of our society's ills. This second, very Malthusian-inspired current sometimes takes an extremist, anti-humanist, even elitist form when it advocates the implementation of brutal eugenics often associated with barely concealed racism. These excesses are well known and dangerous, but we must not throw the baby out with the bathwater. The demographic issue deserves better than that. It is quite possible to envisage population control that does not fall into elitism, eugenics or racism. A population control that responds not only to the requirement to protect the environment on which we depend, but also to that of preserving humanity from a probable collapse. Because there is no secret. Wealth production being dependent on the limited resources of our ecosystem, the share that goes to each individual is necessarily a function of the number of humans. This balance between human well-being and planet preservation requires an enlightened approach to the demographic issue.[12] Without this, for lack of rational population control, we should expect a large-scale demographic collapse, as is typically observed in all species that reproduce to the breaking point in a limited ecosystem. This has already happened in the tragic history of isolated human populations and we would do well to learn from these tragedies that Jared Diamond and others have so finely analysed.

But why is there such strong distrust of the issue of population control? My hypothesis is that the fundamental reason is much deeper than those usually alleged by those who refuse any form of demographic regulation. Monotheistic religions provide part of the explanation for this distrust as they systematically contain the injunction to reproduce and occupy all earthly space. The fierce opposition to contraception and abortion in religious conservative circles is a well-known expression of this. But it is likely that religion does nothing more than formulate, through divine speech, what is most biologically determined in our human behaviour: our fierce and biological obsession with transmitting our genes. The biblical injunction "Be fruitful and multiply" is nothing more than the maxim blindly and implicitly followed by any living organism, from bacteria to humans, including rats and rabbits. If the very idea of population control raises such vehement opposition, it is because it opposes the deep, biological and archaic forces that determine one of the behaviours most anciently inscribed in our psyche: to transmit our genes to our offspring. Opposing it is seen as against nature, as a form of species suicide that consciously renounces a fundamental biological determinism. The particular case of the problem posed by population control is the archetype of the difficulty we have in escaping the biological and psychological determinisms presented in this book: those that have made us grow until we become obese, and obese until we threaten

[11] That said, such a relationship is not linear because humanity produces so much more wealth as it is numerous.

[12] Cazalis and Henderson (2018).

the entire planet. Evolution has endowed us with a brain with such capacities that it is now dangerous to obey the most universal biological determinism there is: to grow and reproduce. We can no longer behave today like rats proliferating without consciousness on the island they have colonised. The realisation of the absurdity of this growth, both economic and demographic, should lead us to free ourselves from the yoke imposed by evolutionary forces. Our survival on Earth, the preservation of our culture, of our civilisation will depend on our ability to ensure that rationality takes precedence over these so pervasive archaisms. And this will not be easy as it is difficult for a human to renounce, defer or moderate the universal desire to have offspring. We touch here on one of the most powerful desires: the one that governs the most fundamental evolutionary mechanisms. It should be noted, however, that the demographic transition is indeed a sign that humans can naturally moderate their propensity to grow. The Ginks phenomenon[13] constitutes a radical form of this, which has indeed sparked violent criticism, precisely because it breaks with a norm whose origin is more biological than rational.

While I believe that reducing our individual consumption is the priority, I do not think we can continue to ignore the demographic issue. In fact, if economic and liberal orthodoxy, as well as most states, see little inconvenience in population growth, quite the contrary, it is primarily because a high birth rate contributes to economic growth. In the world of finance and the most radical liberalism, population growth is necessary for the continued enrichment of a global elite. Because the conjunction of scarcity, reinforced by population growth, ensures further profit growth. Our demographic growth fuels our growth and the interest it presents in the eyes of governments, large companies or Global finance is more of an economic, even geostrategic, order than philosophical or ethical. What is so imperative about the need to grow demographically and what is so monstrously inhuman or philosophically contestable about accepting that reproduction is not an end in itself? More concretely, what difference would it make for humanity to be 4 billion rather than 10 or 15? Is there any reason to prefer that we be 15 billion rather than 10 or 5? Does the expansionism of the species have particular virtues? The whole question remains how to accelerate the demographic transition already underway and how to limit, or even regulate, the human population in the longer term? On this subject, two imperatives must, in my opinion, be respected. Any demographic control objective must be democratically approved and this control must apply identically for all, regardless of course of their level of wealth or social status. This can be perfectly operationalised in the IEIA. There is a cold rationality in considering that bringing a child into the world contributes to accentuating one's own ecological footprint, but it is an undeniable fact. Therefore, if there is no obstacle to each couple deciding whether or not to have a child, it makes sense to integrate, in one way or another, this dimension into the IEIA. It is up to everyone to choose between the immense satisfaction of giving life and that of being able to enjoy the quota of resources

[13] **GINK** is the acronym for *Green Inclination No Kid*, an English term to designate people who make the choice, for ecological reasons, not to have a child.

defined by the IEIA ceiling in a different way. I am well aware of the unnatural nature of this type of alternative. This is only the consequence of the introduction of necessary rationality within the framework of the most fundamental biological process there is: our propensity to reproduce.

10.11 Loss of Freedom?

In his morning column on France Inter, François Morel cleverly ridiculed an advertisement for duck confit that exploited the distrust of many French people towards political correctness and the freedom restrictions that political ecology seems to bring in its punitive component. The text accompanying this advertisement was as follows:

> So we can no longer use our car without feeling guilty, we can no longer hold the door for ladies without being seen as a flirt, soon we will no longer be able to sit on the grass if not we will crush rare insects.

And Morel to continue ironically on a list he improvises from the same stupid rhetoric:

> soon we will no longer be able to kill an elephant when there is nothing prettier than an ivory statuette, soon we will no longer be able to turn on the light at home when we return the next day at the latest, we will no longer be able to eat strawberries in winter on the pretext that it warms the planet, we will no longer be able to throw a plastic bag on the pretext that it will end up in the ocean…

To finally take up the conclusion of this advertisement:

> Fortunately we still have duck confit. A good quick duck confit in the oven without fuss, accompanied by sautéed potatoes with garlic while remaking the world in the kitchen with friends, right!.

By parodying this advertisement, François Morel puts his finger on two essential points. The first concerns the distrust of many citizens towards measures considered punitive, austere and freedom-restricting. The second leads us to identify the absurdity of making individual freedom the cardinal value guiding all political action when, in reality, the banner of freedom is often the alibi that the dominant give themselves to exploit a vast majority of humans.

Regarding the first point, the solution lies almost entirely in education: in the dissemination of information necessary for the acceptance of measures considered punitive—those that could be associated with the implementation of the IEIA—leading to a reduction in our standard of living: a vast programme as the inequality of income and wealth in the world is accompanied by a monumental educational inequality.

Regarding the second point—the problem of freedom—I would like to highlight a theoretical aspect that naturally follows from the argumentative basis of this book. In most cases, the law, and the freedom restrictions it imposes, are safeguards that men have given themselves, first through divine speech and then through institutions, to prevent certain components of human nature from jeopardising the social contract without which any society falls into physical as well as psychological violence. The great revolution that I advocate is that the law should no longer only protect Men from concupiscence, selfishness or violence of Men on other Men. The law should protect nature from the predations that Men naturally exert on it since it behaves towards its ecosystem like almost all other animals if it is not constrained in one way or another. In fact, in this specific case, the law would make us a very original species that is no longer simply the plaything of biological determinisms, but that makes the difficult choice to submit its animality to a rationality without which our survival on the planet is improbable. Inevitably, taking the ecological crisis seriously will lead to restrictions perceived as freedom-restricting since it is obvious that our relationship with the world is a predatory one that needs to be regulated.

This predatory relationship is perfectly universal and it reveals its dangerousness in the case of invasive species accidentally or voluntarily introduced on islands. All animal species tend to grow as much as they can—our obsession with economic growth is just a particular expression of this—and only stop doing so when the overexploitation of their ecosystem forces them to decrease or disappear. None make an exception because we are all, as living beings, driven by a powerful underground force leading us individually to want to reproduce and, collectively, to occupy the available space as densely as possible to exploit it *ad infinitum* its trophic resources.

When a new species is introduced into an island, the overall balance of the ecosystem is altered, which had, prior to this introduction, converged towards a stable state through the complex relationships between all the living organisms that constitute it. This balance corresponds to the natural evolution of any complex system which, over the long course of biological evolution, always converges towards an optimal state and maximum biodiversity. However, due to the limited size of island environments, the intrinsic fragility of their ecosystem and their high rate of endemism, the introduction of invasive species is almost always catastrophic. In general, invasive species and invading species—humans are both—grow exponentially as long as the environment provides them with sufficient trophic resources. But, when the density of the invasive species becomes too high, different forms of regulation intervene and often lead to its demographic collapse. This can be due to epidemics that develop all the more as biodiversity is low. Most often, it is the proliferation of the species that leads to a tipping point such that excessive and uncontrolled consumption of resources leads to a sometimes catastrophic decline in its population. This is obviously what will happen to the human species whose destructive capacities are immeasurably greater than all other invasive or invading species on the planet. To avoid this type of collapse, our species being apparently incapable of behaving differently than rats or goats imported onto islands, there are no other solutions than to simply limit our freedom to consume. A poor reduction in freedom in reality, but a reduction with profound consequences.

But what about more concretely? How would the adoption of the IEIA be freedom-destroying and, even worse, how could it induce a start of totalitarianism? The only deprivation of freedom concerns the freedom to consume as much as one would like. Neither the freedom to think, to express oneself, to create, to exercise one's right to protest are therefore questioned. Nor even the freedom to enrich oneself is questioned for that matter. The constraint consists only in turning away from essentially material consumptions with low added value in terms of personal fulfilment and psychic balance. It may even be the most effective way to refocus on the essentials and free oneself from a dangerous addiction which has the sad privilege not only of harming individuals but especially of harming the environment and others. As for the risk of totalitarianism, it seems to me more theoretical than real as long as we do not wait to be on the edge of the precipice to agree to act. There can be no successful revolutions without the consent of the majority and not that of a single elite. A revolution cannot be anti-democratic if it wants to succeed. The type of revolution that would be brought about by the adoption of the IEIA would considerably improve the lives of a majority of humans and I have no doubt that many would support it if its advantages were explained to them. But, it is obvious that a very powerful minority, most of the citizens of Western countries today, would oppose it. No one wants to change the rules of a game where one wins, even when one knows that these rules are not only unfair, but also, untenable in the long term. All this will lead us to a power struggle but a power struggle of a new kind that will no longer concern the political orientation of a nation, but that of humanity.

10.12 A Delicate Compromise: Deprivation of Freedom and Preservation of the Planet

As soon as one wishes to set up a global political system that tries to free itself from the biological and psychological forces that determine human behaviour, the probability of an authoritarian and freedom-destroying governance increases. Isn't there something inevitably constraining about preventing the strongest from exploiting the weakest? Isn't it freedom-destroying to prevent the enrichment of some even when it is at the expense of others? Human societies cannot completely let express the natural propensity of men to want to dominate others and to exploit without limit the resources of the planet. But they cannot either constrain them excessively authoritatively without generating other types of disorder. The ridge line is narrow. We saw it clearly with the coronavirus episode. Many citizens of all the countries concerned rebelled against health measures that they perceived as freedom-destroying, this perception probably being accentuated both by a lack of information and by the fact that they were adopted too late. The situation would probably have been less conflictual if these measures had been adopted earlier. The problem is exactly the same in terms of ecology, much more serious of course, since it is about the preservation of the planet and our civilisations. The later we admit the need to change our economic-political software, the more we will have to take radical measures to deal with the environmental problem, the greater the risk is that a large-scale authoritarianism will emerge. What is certain is that any realistic and rational ecological policy is necessarily more freedom-destroying than the current economic liberalism. How could it be otherwise since the heart of the ecological crisis lies in the fact that humans are transgressing the limits beyond which the planet is in danger? The freedom-destroying dimension is only the consequence of the need to take into account these limits. Not to recognise it would be as absurd as considering that prohibiting driving in town at more than 100 km/h is an infringement of our freedom. Fortunately, what appears freedom-destroying at one moment in our history, may no longer appear as such at another moment, as soon as the deprivation of freedom appears legitimate to the majority. A few decades ago, hunters paraded around showing off the dozens of vultures they had just killed. I don't know what pleasure they took in this barbarity, but the fact is that they would have considered it liberticidal if they were forbidden this stupid hunt. Today, no one disputes the need to prevent this type of stupid feat and the current regulation no longer seems liberticidal to us, even though it is a bit in the absolute. The liberticidal dimension of any regulation is very relative. Its perception varies with the degree of understanding we have of it. Let's take another example, among so many others, which only moderately offends us today even though it is truly scandalous. Many billionaires own pleasure boats the size of small liners: excess has increased considerably in recent decades. Would it be liberticidal to ban this type of extravagant boating when we know that a day of sailing on these monsters must have a greater ecological impact than that of an average French person over a lifetime? These people contribute on a large scale to the destruction of the planet in terms of carbon emissions and pollution of all kinds and therefore the destruction of biodiversity. What is so fundamental in the freedom to consume that we can allow some of the wealthiest to destroy a common good on which everyone's life depends? What is so precious about the freedom we have to consume as we see fit, even when this material consumption harms others and often ourselves? What there is in reality, beyond the alibi of freedom, is the power of archaic forces, the same ones that lead rats, goats, and other animals introduced on islands to grow blindly until they destroy the ecosystem they depend on. These forces inherent in living beings, these fundamental biological and psychological forces that I have described throughout this book and which lead human to refuse any limitation in his frenzy of consumption. If humans fail to show more awareness and intelligence than rats and goats, if they fail to free themselves from the archaic forces that unconsciously animate them, then the fall will be brutal and in proportion to the damage that we will have inflicted on the planet.

10.13 The Dangerous Tendency Towards the Status Quo

I am absolutely certain that these proposals will not win approval and I anticipate the mocking speeches that will point out the utopia of such a project or those, more violent and sharp, which will stigmatise its liberticidal dimension. That's normal. The violence of the reactions to come, as well as the perception of the utopian nature of these proposals, simply reveal the extent of the journey to be made. I have deliberately chosen to bring out these concrete proposals because it is probably the most effective way to perceive the distance between our society and what would be necessary to put in place to avoid ecological disaster: so many citizens endorse political ecology without really understanding its implications. Because whether we like it or not, there are not thirty-six solutions. There are none other than to limit individual consumption because, ultimately, it is the sum of individual consumptions that is at the origin of the environmental crisis. Humans being what they are, it is very unlikely that they will be able to restrain themselves and a fortiori, that they will consent without assurance that the rest of the planet will do the same.

The proposal I make is therefore utopian in the sense that men are absolutely not ready to abandon an existing system whose origin dates back to the Neolithic. But this proposal is rational and realistic, not only in terms of taking into account the objective—preserve the planet and keep it viable for humanity—but also in terms of the psychological forces that drive us. We must also take into account the fact that we are victims of a bias that I will call the status quo, a bias according to which what is, will be. This bias leads us to think that after a long evolution, the society that human has built is a kind of culmination, a stable state that will only allow minor changes. It's the same phenomenon that leads us to believe that biological evolution has stopped when it never stops. Between human societies 10,000 years ago and our society, the gap is huge. Never a human of the Paleolithic, even the most imaginative, could have conceived the world that is ours. We probably can't imagine the world as it will be in just 100 years because evolution is always accelerating, it is certain that the society of the future will have little in common with ours, not only technologically, but anthropologically. If there is a certain gloom and a great anxiety about our future, it is probably because we feel confusedly that we are living the end of a world, the end of a cycle that has lasted a little over 10,000 years; the end of a society that has based its organisation around the concepts of growth, inequality, competition and technological innovation. Something is not right, a form of relationship to the world and to others that is not sustainable because, even locked in the strongest denial, we know well that we are going to stop growing, willingly or by force. But, the ground suddenly gives way at the mere idea of limiting ourselves in our growth. Faced with the global revolution that this imposes, faced with the void that this brings, faced with an unknown that is worrying by definition, the most reassuring, and yet the least adaptive, behaviour is the status quo, especially since the brain of *Sapiens* is cruelly victim of the tyranny of the present. Between a threat that we perceive as distant and therefore uncertain and a current situation so comfortable for many of us, our brain will for a long time tend to choose the status quo. We must do everything we can to free ourselves from this non-adaptive attitude.

References

Campéon, A. (2011). Ordinary old age in solitude. *Gerontology and Society, 33*, 217–229.

Cazalis, V., Loreau, M., & Henderson, K. (2018). Do we have to choose between feeding the human population and conserving nature? Modelling the global dependence of people on ecosystem services. *Science of the Total Environment, 634*, 1463–1474.

Gifford, R. (2011). The dragons of inaction psychological barriers that limit climate change mitigation and adaptation. *American Psychological Association, 66*(4), 290–302.

Sahlins, M. (1968). *The original affluent society*. (Reprinted in Gowdy, 1998) (pp. 5–41). Routledge.

Sörqvist, P., & Langeborg, L. (2019). Why people harm the environment although they try to treat it well: An evolutionary-cognitive perspective on climate compensation. *Frontiers in Psychology, 10*, 434719. https://doi.org/10.3389/fpsyg.2019.00348

Steffen, W., et al. (2018). Trajectories of the earth system in the anthropocene. *PNAS, 115*(33), 8252–8259.

Van de Velde, C. (2018). Presentation: Sociology of solitude: Concepts, challenges, perspectives. *Sociologie et sociétés, 50*(1), 5–20.

Chapter 11
Beyond Reason, the Beauty of the World

Abstract In the previous chapters, I proceeded in a very academic manner, wishing to be demonstrative in the logical sense of the term in my analysis and the conclusions that arise from it. However, I must admit that the driving force behind this reflection is perhaps primarily poetic and aesthetic. It is the wonder at the beauty of our universe and my empathy for our humanity that motivated this reflection. It is likely that this emotional dimension is necessary for the acceptance of the need to abandon our current way of life, hence the need to allow humans to emotionally reconnect with nature.

My scientific and philosophical education, academic journey, and intellectual sensitivity have led me to propose an argument marked by rationality and critical analysis, all supported by undeniable facts and a philosophical and political stance that I have outlined in the most transparent way possible. Therefore, I wanted this text to be almost demonstrative, and the solution I proposed in the previous chapter should only appear as the logical consequence of premises that can be summarised as follows. Our relationship with the planet is destructive, and it is so due to biological and psychological determinisms from which we cannot individually escape. Consequently, only a limitation of our individual consumption organised by states is likely to preserve an ecosystem on which we are dependent.

The rationalist and analytical orientation that I adopted throughout this work has thus allowed me to dismiss mystical, regressive, and anti-humanist forms of ecological thought, all real obstacles to the revolution that I support. This was necessary to dispel vehement discourse, catch-all slogans, and classic injunctions urgently asking us to stop destroying the environment without these pious wishes being accompanied by a tight analysis of the causes of the current environmental crisis or concrete proposals for its resolution. This resolutely rationalist approach was quite incompatible with a form of expression more underpinned by the emotional

T. Ripoll, *Human Psychology and the Destruction of the Planet*, Anthropocene – Humanities and Social Sciences, https://doi.org/10.1007/978-3-032-02080-2_11

dimension, notably that aroused by simple contact with nature. I would like to rectify this small injustice here, an injustice that is important in terms of our collective ability to act for the environment.

In orthodox argumentative logic, it is appropriate to start from premises to reach a conclusion through reasoning that must be impeccable both in terms of the information on which it is based and the inferences that have allowed us to reach the conclusion. This is what I tried to do in this work. In truth, and despite the apparent rationality of the argumentative form, the initial impulse owes much more to emotion than to reason. It is very classic to start from the conclusion and to search for or develop ex post facto the arguments likely to support it: these are post hoc justifications. Argumentative rationality is ultimately only what we develop afterwards to support a conclusion that initially imposed itself not because of any form of logic, but because of elements as debatable and subjective as ideology, beliefs or emotions: the veneer of rationality cleverly deposited on raw belief or emotion. To be completely honest, this work does not escape this reality. Long before I was an academic and a scientist, long before I methodically explored the fields from which I drew inspiration for this work, I was first and foremost a naturalist: a field man. A child who spent thousands of hours exploring every corner of nature to discover a shiny beetle, a discreet bird, or an endemic flower. A child early on enchanted by the magical flight of a large peacock moth, fascinated by the howling of the wolf, amazed by the infinite richness of small ponds in spring. But also a human being capable of marvelling, of feeling this close link that binds us to nature and of taking immense and deep pleasure in simply watching it live. Finally, and perhaps even more so, an individual who, like others, knows how to look at living things and suffers physically every time the destructive madness of men contributes to its destruction. In this respect, the dynamic that led me to write this book is no different from that which animated Thoreau or Muir, environmental thinkers first sensitive to the beauty of the world before being sensitive to its necessary preservation.

11.1 Can Wonder at Nature Be a Factor Leading Us to Want to Preserve It?

Let's leave this intimate parenthesis and look at its meaning. The motivations that lead humans to act are much less rational than is generally believed. The emotional, affective, hedonic dimension plays a considerable role and it is often much easier to manipulate the behaviour of others by playing on the emotional, even irrational, fibre than by relying on rigorous argumentative logic and undeniable empirical arguments. Great political leaders, especially those at the extremes, or certain radical religious leaders, have understood this for a long time. Individuals are not led to sacrifice their lives on the basis of rational arguments because, for most Sapiens brains, reason weighs relatively little compared to the strength of our beliefs, our affect, our obsession with pleasure. The consequence of all this is that it is unlikely that humanity will agree to change its trajectory on the basis of rational and empirical arguments alone. From this point of view, the ability to marvel at nature, the ability to simply enjoy walking in it, the ability to perceive its richness and to grasp its infinite poetry are certainly a necessary condition for the modification of our behaviours and the acceptance of the frustration that any form of consumption limitation will impose.

What about this emotional dimension, and can we count on it to drive a momentum towards an ecologically compatible human trajectory? The great biologist Edward Wilson, in a book published in 1984, developed the idea that humans would be genetically determined to feel a strong affinity with the natural world. The term biophilia, borrowed from the psychologist Eric Fromm, corresponds to a natural tendency of all humans to love living things, especially species that are close to them. It is a fundamental psychological component that leads us to be moved when we perceive an animal suffering or when a once preserved landscape is disfigured by human action. It is probably also what can generating a feeling of intense emotion when a human locks eyes with a wild animal or a feeling of infinite tranquillity and inner peace in front of a sunrise over the sea. This type of positive emotion has an objective effect on our psychic state as well as our somatic state. Thoreau was not mistaken when he asserted that "No human has ever imagined how much the dialogue with the surrounding nature affected his health or his ailments." His intuition has since been empirically confirmed: proximity to nature contributes to our psychic balance by reducing our level of anxiety, stress, depression.[1] Nature also has a direct impact on our physical health and strongly contributes to our somatic recovery after undergoing an operation or overcoming any disease. This effect is so powerful that the simple act of installing plants in a room helps to accelerate the healing of patients,[2] including for example the speed of healing following an operation. Likewise, walking in the forest is much more beneficial than walking in the city both physiologically and psychologically. Clearly, proximity to nature arouses our wonder as it contributes to our physical and mental well-being. From a certain point of view, this is rather good news revealing that we might perhaps rely on this very strong relationship between human and nature to succeed in modifying our behaviour and making it compatible with the preservation of the planet. This is a thesis that makes sense but I would qualify it for two main reasons.

The first is that this special relationship between human and nature did not wait for our century to exist. It must be as old as human himself and it does not seem to have massively contributed to preserving nature and its biodiversity. How many prehistoric hunters or contemporary hunters strangely associate their wonder at nature and the pleasure of killing the beast they sincerely claim to love? On the positive side, biophilia seems to be gaining strength. There are countless nature protection associations and the very fact that more and more individuals are coming to stop eating meat is undoubtedly the expression of an evolution leading us to respect life more.

The second reason seems more problematic to me. Paradoxically, a source of pleasure or well-being only retains its beneficial effect if it remains accessible and present in one way or another in our daily lives. When it disappears for too long, it

---

[1] Park et al. (2008).

[2] Stefan (2015).

ends up losing its flavour to the point that its loss no longer arouses either desire or frustration. This is the popular meaning of the proverb "out of sight, out of mind". To appreciate something, you must at least occasionally have the opportunity to have a concrete experience of it. Regular practice is often necessary for the pleasure associated with it: we generally stop loving when the loved one is so distant, so evanescent and so virtual that he is no longer able to generate desire. Once the link is broken, the being who mattered so much no longer matters. In the same way, we stop needing to listen to music when we haven't listened to it for a long time; we stop needing to feel our body live when we have abandoned all physical activity for too long... I fear that the evolution of our society is producing exactly the same phenomenon with regard to our relationship with nature. Today, more than half of humans are city dwellers. This will be the case for about 70% of them by 2050 according to a recent UN study. As for the remaining 30%, without living in gigantic megacities, many will live in rural and agricultural environments sometimes just as degraded and devoid of any form of natural life as cities. Among today's urban dwellers, particularly in Africa and Asia, a large proportion have probably never seen a forest, a mountain, the ocean or any spaces allowing them to taste the simple pleasure of being in nature. The misery in which many live is such that they will never have the opportunity to see even a natural space not impacted by the presence of Man... For many city dwellers, nature will have become so abstract, almost so foreign that they will become incapable of grasping its infinite beauty. This phenomenon is already very present even in our country which is still preserved. I meet more and more people who are now unable to enjoy the free pleasure of being in nature and who no longer manage to perceive its infinite richness: the beginnings of spring, the song of a warbler, the tracks of a roe deer, the wonderful flight of cranes or even the wind laden with humidity announcing the passage of a depression. Here they are cut off from nature and ultimately much more comfortable, much more stimulated, much more dependent on urban space. This is now where they feel at home, almost uneasy or impressed by a nature they perceive as austere and sometimes boring. Therefore, how would it be possible to encourage humans to modify their behaviour for an entity they no longer know, they have sometimes never seen and with which they no longer have any relationship? Allowing as many people as possible to reconnect with nature, teaching them to look, to marvel, to feel, to listen, to be in its midst without any other objective than to be in its midst could become a fundamental challenge and an ambitious objective because it is not certain that reason alone will be enough to encourage us to behave differently than we do individually and collectively.

11.2 A Crucial but Probably Badly Engaged Fight

In "The Human Bug" then "Where is the Meaning?", S. Bohler develops an analysis quite close to mine according to which our brain system is caught between, on the one hand, archaic neural structures (the striatum in particular) subjected to the tyranny of the present and in search of immediate satisfactions such as those that generate our addictions and our excesses in terms of consumption and, on the other hand, our frontal cortex, the seat of our consciousness and our most rational analyses, notably those that allow us to project ourselves in the long term and resist these fleeting pleasures that contribute to the destruction of the planet. In this opposition, between frontal cortex and striatum, it is indeed the latter that often takes the upper hand especially as our society does everything to make us slave to the striatum. The latter is so powerful that it can even enslave our frontal system. This is what happens when we try to legitimise and rationalise our least virtuous behaviours in terms of the environment: the alibi of the liberticidal dimension of ecology is high among the pseudo-rational arguments that we develop in order to continue to enjoy all these small material pleasures that lead us to disaster. Despite everything, the discomfort that we often feel in the face of the environmental crisis and the fact that we are in a state of cognitive dissonance result from the fact that even if we give in to the tyranny of the striatum, another part of ourselves, more in relation with the frontal cortex, identifies that our behaviour is neither adapted nor reasonable in the current context of major environmental crisis. The neuronal structure that alerts us in the case where we are in such a state of dissonance is the anterior cingulate cortex. The activity of this neuronal structure is associated with a disturbed and often painful psychic state corresponding to the awareness of behaviours in inadequacy with our values. One way to reduce the activity of this structure is precisely to engage in an activity likely to reduce this state of dissonance. But here, everything and its opposite, the best as well as the worst, remain possible. We can try to reduce our internal tension by consuming even more, by indulging in dire addictions or, on the contrary, by adopting a more virtuous lifestyle compatible with our awareness of the current environmental crisis. S.Bohler is, from this point of view, more optimistic than I am. He thinks that the common goal of all humans to work for the preservation of the planet could constitute the meaning we need to collectively do everything possible to achieve this goal. From this point of view, I think that the almost carnal relationship and wonder towards nature that I mentioned earlier should play a major role because it is perhaps the only one capable of offering us the extra happiness, well-being and meaning capable of matching our addiction to the infinite pleasures resulting from our addiction to consumption. No doubt playing on this aesthetic sensitivity and our natural biophilia is necessary, but I do not believe that this will be enough. Of course, some would be able to modify their behaviour, but others will not be able to for all the reasons presented in this book or will be too late in view of the seriousness of the current situation. Above all, even those who would be ready to live differently, those who would be ready to lose some of their material comfort for the common good, will refuse the "sacrifice" as soon as others do not consent to the same effort. To use Bohler's prose, it is not enough to individually manage to curb one's own striatum; this inner battle victorious of reason over our impulses must occur at the level of humanity as a whole. Because this does not seem to me either possible, or achievable soon enough, I see no other solutions than a minimal form of constraint and control, even if it is of course democratically adopted: this is the meaning of the concrete proposals of the previous chapter.

I am well aware that it is not very opportune to end on a sad or pessimistic note especially as I remain an absurd optimist to use Bruno Latour's expression[3]: I am intimately convinced that we will effectively change our trajectory while being perfectly aware of everything that opposes this inflection. It must be acknowledged, however, that we are far from the revolution that I have outlined even though the urgency is total. I unfortunately believe that it is another type of emotion that will collectively lead men to change their relationship with the world: that linked to the concrete perception of disaster when it will massively and concretely impact the daily life of the majority. The situation is not much different from that of smokers who put an end to their tragic addiction as soon as they are told that they are victims of a cancer that they have a chance to overcome only if they stop smoking. The whole question is to know when this moment will occur because the later it happens the more difficult the transition will be and the more likely the risk of an anthropological regression or a form of global collapse will become. Everyone has to contribute to this revolution to the best of their ability. I hope to have modestly contributed to it by writing this book. It will never compensate for the fact that I have not personally managed to align my ideal and my actions: I am just a poor *Sapiens*.

References

Latour, B. (2021). *Where am I?* Think Round Preventers.

Park, B. J., Tsunetsugu, Y., Ishii, H., Furuhashi, S., Hirano, H., & Kagawa, T. (2008). Physiological effects of Shinrin-yoku (taking in the atmosphere of the forest) in a mixed forest in Shinano Town, Japan? *Scandinavian Journal of Forest Research, 23*(3), 278–283.

Stefan, J. (2015). Influence of indoor and outdoor plants on health: Synthesis of research. *Canadian Psychology, 56*(4), 405–425.

[3] Latour (2021).

Postface

On the brink of the abyss, we would have no choice.

One can applaud an analysis without mechanically accepting its conclusions. This is exactly the feeling I have when reading the important book by Thierry Ripoll. It is the same feeling that I had after finishing *The Human Bug* by Sébastien Bohler who wrote the preface to the text you have in your hands.

I have no doubt about the significant risks we are taking with our species by abandoning it to some of its most ancestral compulsions. This is indeed the theme I return to, albeit in a different way, in my book *Cognitive Apocalypse*. However, I maintain, in the face of the announced climate catastrophe, a certain scepticism about the call for degrowth. This call seems to me to give in to a form of error in reasoning that is quite difficult to discern but which can be typified as like this: since economic growth is accompanied by untenable side effects, notably through the production of CO2 *then* let's eliminate the cause and we won't have to suffer the effects. It's as beautiful as a syllogism but I perceive a paralogistic dimension.

I was able to argue in a previous book *The Planet of Men* (Puf) a point that I would like to briefly recall today: the elimination of negative causes is defensible if it does not produce even more harmful secondary consequences. This is the whole problem posed by Hans Jonas' principle of responsibility that Thierry Ripoll also invokes in his book and which inevitably leads to the idea of degrowth.

This German philosopher, a student of Husserl and Heidegger, became known well beyond philosophical and academic circles by publishing, in 1979 *The Principle of Responsibility*. According to him, humanity has an invaluable value and must be preserved from dangers that could threaten it. However, technological development (genetic engineering, armament, agronomic engineering...) can allow our elimination, and, as such, it must be strictly regulated by norms that Jonas offers to define. For him, those who conceive the technological innovations should, before being allowed to introduce them into our societies, prove that they present no environmental and health risk, either in the present or for future generations. Given the power

T. Ripoll, *Human Psychology and the Destruction of the Planet*, Anthropocene – Humanities and Social Sciences, https://doi.org/10.1007/978-3-032-02080-2

that these new techniques confer on it, it is fundamental, according to the German philosopher, that humanity accepts new principles of responsibility. He particularly insists on the need to renounce any technology that would carry apocalyptic risks. In other words, this means that any technology carrying a risk (even improbable) of destroying humanity must be banned. This is the imperative *in dubio pro malo*: in case of doubt consider the worst. On this subject, Jonas himself used the term "heuristic of fear". This heuristic is supposed to act as an intimidation of a nature to prevent the dangers represented by technosciences. By instilling fear of a terrifying future, Jonas hopes to spread ethical representations that will allow human to avoid the end of times. Here is how Jonas explains it (2000, p.135):

"It results, among other things, in what I call in my reflections the "heuristic of fear": when the principle of hope no longer has inspiring force, then perhaps the warning of fear can lead us to reason. Fear may not in itself be a noble position, but it is quite legitimate. And if there is something to fear, the predisposition to a justified fear is in itself an ethical commandment".

The problem is that the complexity of the contemporary technological world and the natural systems that they can modify is such, that the consequences of our actions become increasingly unpredictable. As we can hardly be assured of the nature of the consequences of our action, continues the promoter of the "heuristic of fear", it is therefore wise to adhere to the maxim *in dubio pro malo*.

This reasoning is criticisable for at least two reasons.

On the one hand, the imagination will always have enough resources to conceive and to make plausible the most pessimistic scenarios, and scientists, faced with extremely complex problems, will always have to concede, if they are honest, that the worst is never strictly impossible. This leads to the fact that certain technological explorations could have been made impossible if this principle had been taken seriously (this is notably the case with the LHC (Large Hadron Collider) as I show in *The Planet of Men*). This implies, moreover, a sensitivity to any proposition based on fear and the imagination of the worst whose success on the market of ideas will not owe everything to the rationality of the arguments that support it. Fear being an excellent media product, we are overwhelmed by incessant health and environmental alerts which are sometimes founded, sometimes not. The uncomfortable result of this is that our contemporaries are afraid, but that dazzled by so many possible fears, they suspend if not their judgement at least their action. This confusion also does not favour a rational editorialisation of the real dangers that we run. Thus, we may well believe that climate change threatens us but not draw many practical conclusions as Thierry Ripoll very well reminds us.

On the other hand, by only considering the possibly catastrophic consequences of our actions we risk not perceiving the no less catastrophic consequences of our inaction. Indeed, by promoting economic degrowth, we mechanically concede technological degrowth (the two are intimately linked as shown by numerous studies) which just as mechanically weakens our possible response to all sorts of dangers that could arise on our collective path. At the time of the pandemic that we are going through, this risk is not hypothetical: it turns out that the climate danger is not the only one that humanity must face and many other probable threats can be easily imagined which can only find responses in the exploration of the possible that science and technology allow. It is indeed from the pharmaceutical industry backed by risk-taking capital, for example, that the vaccine solutions have come, the only hope, at the time I write these lines, to get out of this pandemic crisis.

So *yes* we must think about the uncontrollable emerging effects that our increasingly unregulated appetites produce, however, beware of our immediate intuition that leads us—due to the heuristic of fear—to rush towards the most obvious of solutions (cognitively at least) which would be degrowth.

If we draw all the sap from Jonas's advice—against him if necessary—we will see that it is imprudent to take the risk of interrupting our technological deployment. To Jonas's injunction "Beware that your action does not risk the extinction of life on Earth", we must firmly respond: "Beware that your action *and* your inaction do not risk the extinction of life on Earth".

This is an element all the more crucial as Ripoll, Bohler and I are interested in the social consequences that the ancestral functioning of our brain can produce. It turns out that, precisely, our mind has a lot of trouble making room, especially in the ethical field for the secondary consequences of our actions. A quick example helps to become aware of the problem: Suppose that Jean plays the trifecta, he bets on horse A but is tempted to bet rather on B, which he does not do. The race takes place and it is horse B that wins. If Jean had bet on him, he would have won 10,000 euros…

Paul, for his part, bought a ticket betting on B, but at the last moment he exchanges it with a bettor who bet on horse A following a bad intuition. Since it was horse B who won, he too could have won 10,000 euros.

In your opinion, which of the two will feel the most regret?

There is no need to live these two situations to imagine them quite well. They are both frustrating, but Paul is by far the one who will regret his decision the most. Yet, in both cases, Paul and Jean make a decision that makes them lose 10,000 euros *but* the fundamental difference is that Jean decides *not to act* (that is to say to maintain his bet on A, which is also an action) while Paul *acted* by exchanging his ticket. This is quite revealing of the way our mind works. We are more likely to regret the consequences of an action than those of an inaction. And when it has moral implications, anticipated regrets weigh even heavier. We do not want to be guilty of an action whose consequences would be morally condemnable, we are less careful when these consequences result from inaction. To put it another way, *the immorality of inaction is less perceptible than the immorality of action*. Thus, the consequences of action are generally visible while those of inaction are only supposed.

It seems to me that this is the kind of situation that the injunction to degrowth leads us to: observe the consequences of our actions and confine the consequences of our inaction to the unthinkable. These few lines do not claim to be a solution to the considerable problem posed by the growth/degrowth dilemma. Indeed, I am not even sure that the human being can write the complex models that would allow us to evaluate the value of each societal project. It is therefore with all humility that I concede not knowing, however, what I do know is that I can only be convinced by a mode of demonstration that integrates the primary and secondary consequences of our political projects.

Bibliography

Ainslie, G. (2001). *Breakdown of will*. Cambridge University Press.

Aknin, L. B., Barrington-Leigh, C. P., Dunn, E. W., Helliwell, J. F., Burns, J., Biswas-Diener, R., Kemeza, I., Nyende, P., Ashton-James, C. E., & Norton, M. I. (2013). Prosocial spending and well-being: Crosscultural evidence for a psychological universal. *Journal of Personality and Social Psychology, 104*(4), 635–652.

Alberto, M., & Ravat, J. (2011). *Human morality and sciences*. Editions Matériologiques.

Alexander, A., & Yacoumis, P. (2018). Degrowth, energy descent, and 'low-tech' living: Potential pathways for increased resilience in times of crisis. *Journal of Cleaner Production, 197*, 1840–1848.

Allen, M. W., & Jones, T. L. (2014). *Violence and warfare among hunter-gatherers*. Left Coast Press.

Altemeyer, B. (1998). The other "authoritarian personality". In M. Zanna (Ed.), *Advances in experimental social psychology* (Vol. 30, pp. 47–92). Academic Press.

Altemeyer, B. (2003). What happens when authoritarians inherit the earth: A simulation. *Analyses of Social Issues and Public Policy, 3*, 161–169.

Altemeyer, B. (2004). Highly dominating, highly authoritarian personalities. *The Journal of Social Psychology, 144*(4), 421–447.

Arendt, H. (1959). *The human condition*. University of Chicago Press.

Asseng, S., Foster, I., & Turner, N. (2011). The impact of temperature variability on wheat yields. *Global Change Biology, 17*, 997. https://doi.org/10.1111/j.1365-2486.2010.02262.x

Bala, G., Caldeira, K., Mirin, A., Wickett, M., & Delire, C. (2005). Multicentury changes in global climate and carbon cycle: Results from a coupled climate and carbon model. *Journal of Climate, 18*, 4531–4544.

Barbaro, A., & Earp, J. (2008). *Consuming kids: The commercialisation of childhood*. DVD produced by the Media Education Foundation. http://www.mediaed.org/assets/products/134/presskit_134.pdf

Barclay, L. J., & Saldanha, M. F. (2015). Recovering from organisational injustice: New directions in theory and research. In M. Ambrose & R. Cropanzano (Eds.), *The Oxford handbook of justice in work organisations* (pp. 497–522). Oxford University Press.

Barnosky, A. D., et al. (2012). Approaching a state shift in earth's biosphere. *Nature, 486*, 52–58.

Barrett, J.-L. (2007). Cognitive science of religion: What is it and why is it? *Religion Compass, 1*(6), 768–786.

T. Ripoll, *Human Psychology and the Destruction of the Planet*, Anthropocene – Humanities and Social Sciences, https://doi.org/10.1007/978-3-032-02080-2

Batson, C. D., & Moran, T. (1999). Empathy-induced altruism in a prisoner's dilemma. *European Journal of Social Psychology, 7*, 909–924.

Battisti, D., & Naylor, R. (2009). Historical warnings of future food insecurity with unprecedented seasonal heat. *Science, 323*, 240–244.

Baumeister, R. F., Sparks, E. A., Stillman, T. F., & Vohs, K. D. (2008). Free will in consumer behaviour: Self-control, ego depletion, and choice. *Journal of Consumer Psychology, 18*, 4–13.

Belk, R. W., Lizger, G., & Askegaard, S. (2003). The fire of desire: A multisited inquiry into consumer passion. *Journal of Consumer Research, 30*, 326–351.

Benjamin, B. D., Heffetz, O., Kimball, M. S., & Rees-Jones, A. (2012). What do you think would make you happier? What do you think you would choose? *American Economic Review, 102*(5), 2083–2110.

Bering, J. (2006). The folk psychology of souls. *Behavioural and Brain Sciences, 29*, 453–498.

Berndt, R. M., & Berndt, C. H. (1964). *The world of the first Australians*. Angus & Robertson.

Biermeyer, P. (2001). *Coming changes in the U.S. clothes washer market*. Lawrence Berkeley National Laboratory. Unpublished paper.

Biraben, J.-N. (2003). The rising numbers of humankind. *Population & Societies, 394*, 1–4.

Bjørn, A., Kalbar, P., Elsborg Nygaard, S., Kabins, S., Jensen, C. L., Birkved, M., Schmidt, J., & Hauschild, M. Z. (2018). Pursuing necessary reductions in embedded GHG emissions of developed nations: Will efficiency improvements and changes in consumption get us there? *Global Environmental Change, 52*, 314–324.

Bloom, P. (2004). *Descartes' baby*. Basic Books.

Boas, F. (1920). The social organisation of Kwakiutl. *American Anthropologist, 22*, 111–126.

Bocquet-Appel, J.-P. (2011). When the world's population took off: The springboard of the Neolithic demographic transition. *Science, 333*, 560–561.

Boehm, C. (1997). Impact of the human egalitarian syndrome on Darwinian selection mechanics. *American Naturalist, 150*, 100–121.

Boehm, C. (1999). *Hierarchy in the forest*. Harvard University Press.

Boehm, C. (2008). Purposive social selection and the evolution of human altruism. *Cross-Cultural Research, 42*, 319–352.

Boehm, C. (2012). *Moral origins: The evolution of virtue, altruism, and shame*. Basic Books.

Boehm, C. (2013). The biocultural evolution of conflict resolution between groups. In D. P. Fry (Ed.), *War, peace, and human nature* (pp. 315–340). Oxford University Press.

Bohler, S. (2019). *Le bug humain*. Paris, Robert Laffont.

Bohler, S. (2021). *Où est le sens ? les découvertes sur notre cerveau qui changent l'avenir de notre civilisation*. Paris, Pocket.

Boulding, K. E. (1966). The economics of the coming spaceship earth. In H. Jarrett (Ed.), *Environmental quality in a growing economy*. Resources for the Future/Johns Hopkins University Press.

Bourg, D. (2019). Is ecology authoritarian, even fascist? Or decidedly democratic? *Ecological Thought, 3*(2), 1–13.

Bourguignon, F. (2012). *The globalisation of inequality*. Threshold.

Bowerman, T. (2014). How much is too much? A public opinion research perspective. *Sustainability: Science, Practice, & Policy, 10*(1), 14–28.

Bowles, S. (2006). Group competition, reproductive leveling, and the evolution of human altruism. *Science, 314*, 1569–1572.

Bowles, S. (2011). Cultivation of cereals by the first farmers was not more productive than foraging. *Proceedings of the National Academy of Science, 108*(12), 4760–4765.

Bowles, S., Choi, J.-K., & Hopfensitz, A. (2003). The coevolution of individual behaviours and group level institutions. *Journal of Theoretical Biology, 223*(2), 135–147.

Brickman, P., Coates, D., & Janoff-Bulman, R. (1978). Lottery winners and accident victims: Is happiness relative? *Journal of Personality and Social Psychology, 36*, 917–927.

Bronner, G. (2021). *L'Apocalypse cognitive*. Paris, PUF.

Brown, J. H., Burnside, W. R., Davidson, A. D., Delong, J. P., Dunn, W. C., Hamilton, M. J., Mercado-Silva, N., Nekola, J. C., Okie, J. G., Woodruff, W. H., & Zuo, W. (2011). Energetic limits to economic growth. *Bio Science, 61*, 19–26.

Büchs, M., & Koch, M. (2019). Challenges for the degrowth transition: The debate about wellbeing. *Futures, 105*, 155–165.

Burch-Brow, J., & Alfred Archer, A. (2017). In defence of biodiversity. *Biology and Philosophy, 32*, 969–997.

Busch, R. C. (1990). *Family systems: Comparative study of the family*. P. Lang.

Byrne, R. W., & Whiten, A. (1988). *Machiavellian intelligence: Social expertise and the evolution of intellect in monkeys, apes, and humans*. Clarendon Press/Oxford University Press.

Campbell, W. K., Bush, C. P., Brunell, A. B., & Shelton, J. (2005). Understanding the social costs of narcissism: The case of the tragedy of the commons. *Personality and Social Psychology Bulletin, 31*(10), 1358–1368.

Campéon, A. (2011). Ordinary old age in solitude. *Gerontology and Society, 33*, 217–229.

Carlsson, F. (2007). *Do you enjoy having more than others? Survey evidence of positional*. Working papers in Economics no 100.

Caunt, B. S., Franklin, J., Brodaty, N. E., & Brodaty, H. (2013). Exploring the causes of subjective well-being: A content analysis of peoples' recipes for long-term happiness. *Journal of Happiness Studies, 14*, 475–499.

Cazalis, V., Loreau, M., & Henderson, K. (2018). Do we have to choose between feeding the human population and conserving nature? Modelling the global dependence of people on ecosystem services. *Science of the Total Environment, 634*, 1463–1474.

Chagnon, N. (1977). *Yanomamo: The fierce people* (2nd ed.). Holt.

Chakrabarty, D. (2018). The seventh history and theory lecture anthropocene time. *History and Theory, 57*(1), 5–32.

Choi, J.-K., & Bowles, S. (2007). The coevolution of parochial altruism and war. *Science, 318*(26), 636–640.

Cohen, M., & Crane-Kramer, G. (2007). *Ancient health: Skeletal indicators of agricultural and economic intensification*. University Press of Florida; Patricia Lambert. Health.

Colyvan, M., Linquist, S., Grey, W., Griffiths, P., Odenbaugh, J., & Possingham, H. P. (2009). Philosophical issues in ecology: Recent trends and future directions. *Ecology and Society, 14*(2), 22.

Cook, J., & Jacobs, P. (2014). Scientists are from Mars, laypeople are from Venus: An evidence-based rationale for communicating the consensus on climate. *Reports of the National Center for Science Education, 34*, 3.1–3.10.

Cook, J., Nuccitelli, D., Green, S. A., Richardson, M., Winkler, B., Painting, R., Way, R., Jacobs, P., & Skuce, A. (2013). Quantifying the consensus on anthropogenic global warming in the scientific literature. *Environmental Research Letters, 8*(2), 024024.

Costanza, R., Daly, L., Fioramonti, L., Giovannini, E., Kubiszewki, I., Mortensen, L. F., Pickett, K. E., Ragnarsdottir, K. V., De Vogli, R., & Wilkinson, R. (2016). Modelling and measuring sustainable wellbeing in connection with the UN Sustainable Development Goals. *Ecological Economics, 130*, 350–355.

Costanza, R., Kubiszewski, I., Giovannini, E., Lovins, H., McGlade, J., Pickett, K. E., Ragnarsdóttir, K. V., Roberts, D., De Vogli, R., & Wilkinson, R. (2014). Time to leave GDP behind. *Nature, 505*, 283–285.

Costanza, R., McGlade, J., Lovins, H., & Kubiszewski, I. (2014). An overarching goal for the UN sustainable development goals. *Solutions, 5*(4), 13.

Crist, E. (2007). Beyond the climate crisis: A critique of climate change discourse. *Telos, 141*, 29–55.

Crooks, K. R., & Soule, M. E. (1999). Mesopredator release and avifaunal extinctions in a fragmented system. *Nature, 400*, 563–566.

Dawkins, R. (2001). *Sustainability doesn't come naturally: A Darwinian perspective on values*. The Environment Foundation Inaugural Lecture.

De La Bruslerie, H. (2012). The psychological value of time: A synthesis of the literature. *Economic News, 88*(3), 361–400.

De Neve, J. E., Ward, G. W., De Keulenaer, F., Van Landeghem, W., Kavetsos, G., & Norton, M. I. (2015). *The asymmetric experience of positive and N, G., negative economic growth: Global evidence using subjective well-being data.*, IZA Discussion Papers, No. 8914. Institute for the Study of Labour (IZA).

de Waal, F. B. (2008). Putting the altruism back into altruism: The evolution of empathy. *Annual Review of Psychology, 59*, 279–300.

Denault, L. K., & McFarlane, D. A. (1995). Reciprocal altruism between male vampire bats, Desmodus rotundus. *Animal Behaviour, 49*, 855–856.

Derex, M., Beugin, M. P., & Godelle, B. (2013). Experimental evidence for the influence of group size on cultural complexity. *Nature, 503*, 389–391.

Diamond, J. (1978). The Tasmanians—The longest isolation, the simplest technology. *Nature, 273*, 185–186.

Diamond, J. (2005). *Collapse: How societies choose to fail or succeed.* Viking.

Diener, E., & Seligman, M. E. (2002). Very happy people. *Psychological Science, 13*, 81–84.

Diener, E., Tay, L., & Oishi, S. (2013). Rising income and the subjective well-being of nations. *Journal of Personality & Social Psychology, 104*, 267–276.

Diener, E., et al. (2018). Advances and open questions in the science of subjective well-being. *Collabra: Psychology, 4*(1), 15. https://doi.org/10.1525/collabra

Divale, W. T., & Harris, M. (1976). Population, warfare, and the male supremacist complex. *American Anthropologist, 78*, 521–538.

Dortier, J.-F. (2015). *Revolution in our origins*, under the direction of Jean-François Dortier. Editions, Human Sciences.

Downey, S. S., Randall Haas, A., & Shennan, S. J. (2016). European Neolithic societies showed early warning signals of population collapse. *Proceedings of the National Academy of Sciences, 113*(35), 9751.

Drews, S., & van den Bergh, J. C. J. M. (2017). Scientists' views on economic growth versus the environment: A questionnaire survey among economists and non-economists. *Global Environmental Change, 46*, 88–103.

Duckitt, J. (2001). A dual-process cognitive-motivational theory of ideology and prejudice. *Advances in Experimental Social Psychology, 33*, 41–113.

Dunbar, R. (1998). The social brain hypothesis. *Evolutionary Anthropology, 6*, 178–190.

Dunbar, R., & Shultz, S. (2007). Understanding primate brain evolution. *Philosophical Transactions of the Royal Society B, 362*, 649–658.

Dunlap, R. E., & Van Liere, K. D. (1984). Commitment to the dominant social paradigm and concern for environmental quality. *Social Science Quarterly, 65*, 1013–1028.

Dunn, E. W., Aknin, L. B., & Norton, M. I. (2008). Spending money on others promotes happiness. *Science, 319*, 1687–1688.

Dunn, M. J., & Searle, R. (2010). Effect of manipulated prestige-car ownership on both sex attractiveness ratings. *British Journal of Psychology, 101*(Pt 1), 69–80.

Durand, F. (2018). Global warming: Colossal stakes for new societal projects. *Parcours.* Cahiers du GREP Midi-Pyrénées, GREP MP. hal-02004842.

Durrheim, K., Jacobs, N., & Dixon, J. (2014). Explaining the paradoxical effects of intergroup contact: Paternalistic relations and system justification in domestic labour in South Africa. *International Journal of Intercultural Relations, 41*, 150–164.

Easterlin, R. A. (1974). Does economic growth improve the human lot? Some empirical evidence. *Nations and Households in Economic Growth, 89*, 89–125.

Easterlin, R. A. (1995). Will raising the incomes of all increase the happiness of all? *Journal of Economic Behavior and Organization, 27*, 35–47.

Easterlin, R. A. (2017). Paradox lost? *Review of Behavioural Economics, 4*(4), 311–339.

Easterlin, R. A., McVey, L. A., Switek, M., Sawangfa, O., & Smith Zweig, J. (2010). The happiness–income paradox revisited. *PNAS, 107*(52), 22463–22468.

Ehrlich, P. R. (2000). *Human natures: Genes, cultures, and the human prospect*. Island Press.
Elkins, S. (1967). Slavery and personality. In R. S. Lazarus & E. M. Opton Jr. (Eds.), *Personality* (pp. 395–420). Penguin.
Ellemers, N. (2012). The group self. *Science, 336*, 876–879.
Ellis, E. C. (2015). Ecology in an anthropogenic biosphere. *Ecological Monographs, 85*(3), 287–331.
Ellis, E. C., Goldewijk, K. K., Siebert, S., Lightman, D., & Ramankutty, N. (2010). Anthropogenic transformation of the biomes, 1700 to 2000. *Global Ecology and Biogeography, 19*, 589–606.
Fear, J. (2008). *Stuff happens: Unused things cluttering up our homes.*, Research Paper No. 52. The Australia Institute.
Fehr, E., & Gachter, S. (2002). Altruistic punishment in humans. *Nature, 415*, 137–140.
Ferry, L. (1992). *The new ecological order*. Grasset.
Feygina, I. (2013). Social justice and the human–environment relationship: Common systemic, ideological, and psychological roots and processes. *Social Justice Research, 26*, 363–381.
Feygina, I., Jost, J. T., & Goldsmith, R. E. (2010). System justification, the denial of global warming, and the possibility of "system-sanctioned change". *Personality and Social Psychology Bulletin, 36*, 326–338.
Fioramonti, L. (2013). *Gross domestic problem: The politics behind the world's most powerful number*. Zed Books.
Fittkau, E., & Klinge, H. (1973). On biomass and tropic structure of the central Amazonian rain forest ecosystem. *Biotropica, 5*, 2–14.
Flipo, F. (2010). *Deep ecology, a threatening fundamentalism or a non-modern liberalism?* Cahiers Sens Public Assoc. Sens-Public.
Frischmann, B. M., Marciano, A., & Battista Ramello, B. (2019). Retrospectives tragedy of the commons after 50 years? *The Journal of Economic Perspectives, 33*(4), 211–228.
Gat, A. (2006). *War in human civilisation*. Oxford University Press.
Gat, A. (2015). Proving communal warfare among hunter-gatherers: The quasi-rousseauan error. *Evolutionary Anthropology, 24*, 111–126.
Gavrilets, S., & Fortunato, L. A. (2014). Solution to the collective action problem in between-group conflict with within-group inequality. *Nature Communications, 5*, 3526. https://doi.org/10.1038/ncomms4526
Geary, D., & Bailey, D. (2009). Hominid brain evolution: Testing climatic, ecological, and social competition models. *Human Nature, 20*, 265–279.
Ger, G. L., & Belk, R. (1996). I'd like to buy the world a coke: Consumptionscapes of the 'less affluent world'. *Journal of Consumer Policy, 19*(3), 1–34.
Gervais, W. M., & Norenzayan, A. (2012). Analytic thinking promotes religious disbelief. *Science, 336*, 493–496.
Gifford, R. (2011). The dragons of inaction psychological barriers that limit climate change mitigation and adaptation. *American Psychological Association, 66*(4), 290–302.
Glikson, A. Y., & Groves, C. (2016). *Climate, fire and human evolution: The deep time dimensions of the anthropocene*. London Springer.
Gowdy, J. (Ed.). (1998). *Limited wants, unlimited means: A reader on hunter-gatherer economics and the environment*. Island Press.
Gowdy, J. (2020). Our hunter-gatherer future: Climate change, agriculture and uncivilisation. *Futures, 115*, 1–9.
Gowdy, J., & Krall, L. (2013). The ultrasocial origins of the Anthropocene. *Ecological Economics, 95*, 137–147.
Gowdy, J., & Krall, L. (2014). The transition to agriculture and the evolution of human ultrasociality. *Journal of Bioeconomics, 16*(2), 179–202.
Gowdy, J., & Krall, L. (2016). The economic origins of ultrasociality. *The Behavioural and Brain Sciences, 39*, 1–60.
Griffin, D., & Tversky, A. (1992). The weighing of evidence and the determinants of confidence. *Cognitive Psychology, 24*, 411–435.

Gsottbauer, E., & van den Bergh, J. C. J. M. (2010). Environmental policy theory given bounded rationality and other-regarding preferences. *Environmental & Resource Economics, 49*(2), 263–304.

Hamilton, W. D. (1964). The genetical evolution of social behaviour. I & II. *Journal of Theoretical Biology, 7*(1), 1–52.

Hamilton, C. (2010). *Requiem for a species*. Earthscan.

Hamilton, C., Denniss, R., & Baker, D. (2005). *Wasteful consumption in Australia*. The Australia Institute. Discussion Paper 77.

Hannelore De Silva, H., Hauert, C., Traulsen, A., & Sigmund, K. (2010). Freedom, enforcement, and the social dilemma of strong altruism. *Journal of Evolutionary Economics, 20*, 203–217.

Hardin, G. (1968). The tragedy of the commons. *Science, 162*(3859), 1243–1248.

Harris, C., & Laibson, D. (2001). Dynamic choices of hyperbolic consumers. *Econometrica, 69*(4), 935–957.

Haushofer, J., & Shapiro, J. (2015). *Household response to income changes: Evidence from an unconditional cash transfer program in Kenya*. Working paper. The Busara Center for Behavioural Economics.

Hayden, B. (2008). *Man and inequality. The invention of hierarchy in prehistory*. CNRS Editions.

Heather, N., & Segal, G. (2013). Understanding addiction: Donald Davidson and the problem of akrasia. *Addiction Research and Theory, 21*, 445–452.

Henden, E. (2012). Addictive actions. *Philosophical Psychology, 26*, 1–21. https://doi.org/10.1080/09515089.2012.660688

Henden, E., Melberg, H. O., & Røgeberg, O. J. (2013). Addiction: Choice or compulsion? *Frontiers in Psychiatry, 4*(77), 1–11.

Hennes, E. P., Ruisch, B. C., Feygina, I., Monteiro, C. A., & Jost, J. T. (2016). Motivated recall in the service of the economic system: The case of anthropogenic climate change. *Journal of Experimental Psychology. General, 145*, 755–771.

Henrich, J. (2004). Cultural group selection, coevolutionary processes and large-scale cooperation. *Journal of Economic Behaviour and Organisation, 53*(1), 3–35.

Henrich, J., & Boyd, R. (2002). On modelling cognition and culture: Why cultural evolution does not require replication of representations. *Journal of Cognition and Culture, 2*, 87–112.

Hilt, B. M., & Formicola, V. (2008). Hunters of the ice age: The biology of upper paleolithic people. *Yearbook of Physical Anthropology, 1*, 70–99.

Hsee, C. K., & Hastie, R. (2006). Decision and experience: Why we don't choose what makes us happy? *Trends in Cognitive Sciences, 10*, 31–37.

Hutchings, J. A., & Reynolds, J. D. (2004). Marine fish population collapses: Consequences for recovery and extinction. *Bioscience, 54*, 297–309.

Jackson, T. (2009). *Prosperity without growth: Economics for a finite planet*. Earthscan.

Jackson, T. (2011). Societal transformations for a sustainable economy. *Natural Resources Forum, 35*, 155–164.

Jackson, T., & Marks, N. (1999). Consumption, sustainable well-being and human needs—With reference to UK expenditure patterns 1954–1994. *Ecological Economics, 28*(3), 421–442.

John Gowdy, J., & Baveye, P. (2019). An evolutionary perspective on industrial and sustainable agriculture. In G. Lemaire (Ed.), *Agro-ecosystem diversity: Reconciling contemporary agriculture and environment quality*. Elsevier.

Jonas, H. (1979). *The principle of responsibility, an ethic for technological civilisation*. Cerf.

Jones, S. (1993). *The language of the genes*. Anchor Books.

Jouventin, P. (2016). *Man, this failed animal*. Free & Solidarity.

Kahneman, D. (2000). Experienced utility and objective happiness: A moment-based approach. In D. Kahneman & A. Tversky (Eds.), *Choices, values and frames* (pp. 673–692). Cambridge University Press.

Kahneman, D. (2011). *Thinking, fast and slow*. Penguin.

Kahneman, D., & Deaton, A. (2010). High income improves evaluation of life but not emotional well-being. *PNAS, 107*(38), 16489–16493.

Kaiser, C., & Vendrik, M. (2018). *Different versions of the Easterlin Paradox: New evidence for European countries.*, IZA Discussion Papers, No. 11994. Institute of Labor Economics (IZA).

Kallis, G. (2011). In defence of degrowth. *Ecological Economics, 70*, 873–880.

Kallis, G., et al. (2018). On degrowth. *Annual Review of Environment and Resources, 43*, 4.1–4.26.

Kanheman, D., & Tverski, A. (1979). Prospect theory: An analysis of decisions under risk. *Econometrica, 47*(2), 263–291.

Kaplan, H., Lancaster, K., & Hurtado, M. (2000). A theory of human life history evolution: Diet, intelligence, and longevity. *Evolutionary Anthropology, 9*, 156–185.

Kasting, J. (1998). The carbon cycle, climate, and the long-term effects of fossil fuel burning. *Consequences, 4*, 15–27.

Kelemen, D., & Rosset, R. (2009). The human function compunction: Teleological explanation in adults. *Cognition, 111*, 138–143.

Kerschner, C. (2010). Economic de-growth vs. steady-state economy. *Journal of Cleaner Production, 18*, 544–551.

Knauft, D. (1994). Culture and cooperation in human evolution. In L. Sponsel & T. Gregor (Eds.), *The anthropology of peace and nonviolence* (pp. 37–67). Rienner.

Koomey, J. G., Berard, S., Sanchez, M., & Wong, H. (2011). Implications of historical trends in the electrical efficiency of computing. *Annals of the History of Computing, IEEE, 33*(3), 46–54.

Kubiszewski, I., Costanza, R., Franco, C., Lawn, P., Talberth, J., Jackson, T., & CAylmer, C. (2013). Beyond GDP: Measuring and achieving global genuine progress. *Ecological Economics, 93*, 57–68.

Lane, T. (2017). How does happiness relate to economic behaviour? A review of the literature. *Journal of Behavioral and Experimental Economics, 68*(2017), 62–78.

Larsen, C. S. (2006). The agricultural revolution as environmental catastrophe: Implications for health and lifestyles in the Holocene. *Quaternary International, 150*, 12–20.

Latour, B. (2021). *Where am I?* Think Round Preventers.

Layard, R., Mayraz, G., & Nickell, S. (2010). Does relative income matter? Are the critics right? In E. Diener, J. Helliwell, & D. Kahneman (Eds.), *International differences in well-being*. Oxford University Press.

Lerner, M. J. (1980). *The belief in a just world: A fundamental delusion*. Plenum Press.

Li, L., & Shi, L. (2019). Economic growth and subjective well-being: Analysing the formative mechanism of Easterlin Paradox. *The Journal of Chinese Sociology, 6*, 1. https://doi.org/10.1186/s40711-018-0090-9

Loss, S. R., Will, T., & Marra, P. (2012). Direct human-caused mortality of birds: Improving quantification of magnitude and assessment of population impacts. *Frontiers in Ecology and the Environment, 10*, 357–356.

Lucas, R. E., Clark, A. E., Georgellis, Y., & Diener, E. (2004). Unemployment alters the set point for life satisfaction. *Psychological Science, 15*, 8–13.

Lucas, R., & Donnellan, M. (2007). How stable is happiness? Using the STARTS model to estimate the stability of life satisfaction. *Journal of Research in Personality, 41*(5), 1091–1098.

Lumley, A., Michalczyk, L., & Kitson, J. (2015). Sexual selection protects against extinction. *Nature, 522*, 470–473.

Luttmer, E. F. (2005). Neighbours as negatives: Relative earnings and well-being. *Quarterly Journal of Economics, 120*, 963–1002.

Mace, et al. (2005). Biodiversity, Ch. 4. In H. Hassan, R. Scholes, & N. Ash (Eds.), *Ecosystems and human wellbeing: Current state and trends* (pp. 79–115). Island Press.

Majolo, B. (2019). Warfare in an evolutionary perspective. *Evolutionary Anthropology, 28*, 321–331.

Markey, P., & Markey, C. (2011). Pornography-seeking behaviours following midterm political elections in the United States: A replication of the challenge hypothesis. *Computers in Human Behaviour, 27*, 1262–1264.

Medvec, V. H., Madey, S. F., & Gilovich, T. (1995). When less is more: Counterfactual thinking and satisfaction among Olympic medalists. *Journal of Personality and Social Psychology, 69*, 603–610.

Meloni, M. (2014). How biology became social, and what it means for social theory. *The Sociological Review, 62*, 593–614.

Metzner, R. (1994). The emerging ecological worldview. In M. E. Tucker & J. A. Grim (Eds.), *Worldviews and ecology: Religion, philosophy, and the environment* (pp. 163–172). Orbis Books/Maryknoll.

Milanovic, B., Lindert, P., & Williamson, J. (2007). *Measuring ancient inequality*. World Bank Policy Research Working Paper 4412.

Moriarty, P., & Honnery, D. (2012). What is the global potential for renewable energy? *Renewable and Sustainable Energy Reviews, 16*(1), 244–252.

Murtaugh, P. A., & Schlax, M. G. (2009). Reproduction and the carbon legacies of individuals. *Global Environmental Change, 19*, 14–20.

Myong-Il, K., & Shinsuke, I. (2016). Time discounting, present biases, and health-related behaviours: Evidence from Japan. *Economics and Human Biology, 2*(1), 22–136.

Naess, A. (1973). The shallow and the deep, long range ecology movements. *Inquiry, 16*, 95–100.

Nekola, J. C., Allen, C. D., Brown, H. J., Burger, J. R., Davidson, A. D., Fristoe, T. S., Hamilton, M. J., Hammond, S. T., Kodric-Brown, A., Mercado-Silva, N., & Okie, J. K. (2017). The Malthusian–Darwinian dynamic and the trajectory of civilisation. *Trends in Ecology & Evolution, 28*(3), 128–130.

Oishi, S., Schimmack, U., & Diener, E. (2012). Progressive taxation and the subjective well-being of nations. *Psychological Science, 23*(1), 86–92.

Ojalaa, A., Korpelab, K., Lankic, T., Tsunetsugud, Y., & Kagaw, T. (2014). The influence of urban green environments on stress relief measures: A field experiment. *Journal of Environmental Psychology, 38*, 1–9.

Okin, G. S. (2017). Environmental impacts of food consumption by dogs and cats. *PLoS One, 12*, e0181301. https://doi.org/10.1371/journal.pone

Ostrom, E. (1990). *Governing the commons: The evolution of institutions for collective action (Political economy of institutions and decisions)*. Cambridge University Press.

Park, B. J., Tsunetsugu, Y., Ishii, H., Furuhashi, S., Hirano, H., & Kagawa, T. (2008). Physiological effects of Shinrin-yoku (taking in the atmosphere of the forest) in a mixed forest in Shinano Town, Japan? *Scandinavian Journal of Forest Research, 23*(3), 278–283.

Pathou-Mathis, M. (1993). Paris, Odile Jacob.

Pickett, K., & Wilkinson, R. (2019). *To live happily, let's live equal!* The Bonds that Liberate.

Piketty, T. (2013). *Capital in the twenty-first century paperback*. Le Seuil.

Pinker, S. (2012). *The better angels of our nature. A history of violence and humanity*. Penguin.

Ponting, C. (2007). *A new green history of the world*. Penguin Books.

Pratto, F. (1999). The puzzle of continuing group inequality: Piecing together psychological, social, and cultural forces in social dominance theory. In M. P. Zanna (Ed.), *Advances in experimental social psychology* (Vol. 31, pp. 191–263). Academic Press.

Pringle, H. (2013). The origins of creativity. *Scientific American, 308*(3), 37–43.

Pronin, E. (2007). Perception and misperception of bias in human judgment. *Trends in Cognitive Sciences, 11*(1), 37–43.

Puffer, S. (1999). CompUSA's James Halpin on technology, rewards, and commitment. *Academy of Management Executive, 13*, 29–37.

Pyysiainen, I., & Hauser, M. (2010). The origins of religion: Evolved adaptation or by-product? *Trends in Cognitive Sciences, 14*(3), 104–109.

Ramonet, I. (1996). *New powers, new masters of the world: The great conferences*. Fidès.

Rampell, C. (2009). Shift from spending to saving may be slump's lasting impact. *New York Times*, 10 May 2009.

Raworth, K. (2017). *Doughnut economics: Seven ways to think like a 21st-century economist*. Random House Business.

Rees, J. H., Klug, S., & Bamberg, S. (2015). Guilty conscience: Motivating pro-environmental behaviour by inducing negative moral emotions. *Climat Change, 130*, 439–452.

Richerson, P. J., & Boyd, R. (2005). *Not by genes alone: How culture transformed human evolution*. University of Chicago Press.

Ripoll, T. (2020). *Pourquoi croit-on ? Psychologie des croyances*. Auxerre, Sciences Humaines.

Riveros, A., Seid, M., & Wcislo, W. (2012). Evolution of brain size in class-based societies of fungus-growing ants (Attini). *Animal Behaviour, 83*, 1043–1049.

Robson, A. J. (2002). Evolution of human nature. *Journal of Economic Perspectives-Volume, 16*(2), 89–106.

Rockström, J. (2009). A safe operating space for humanity. *Nature, 461*, 472–475.

Ropke, I. (1999). Theories of practice—New inspiration for ecological economic studies on consumption. *Ecological Economics, 68*, 2490–2497.

Rowland, W. (2009). Reflections on metaphor and identity in the cyber-corporation. *Journal of Business Ethics, 90*, 15–28.

Ryan, R. M., & Deci, E. L. (2001). On happiness and human potentials: A review of research on hedonic and eudaimonic well-being. *Annual Review of Psychology, 52*, 141–166.

Saad, G., & Vongas, J. G. (2009). The effect of conspicuous consumption on men's testosterone levels. *Organizational Behavior and Human Decision Processes, 110*, 80–92.

Sahlins, M. (1968). *The original affluent society*. (Reprinted in Gowdy, 1998) (pp. 5–41). Routledge.

Sajaloli, B., & Grésillon, E. (2016). The Catholic Church, ecology and the protection of the environment: chronicle of a theological and political conversion. *Géoconfluences* (posted online on 19 October 2016).

Samuelson, R. (2005). Homes as hummers. *Washington Post*, 13 July 2005.

Sanchez, P. (2007). *The rationality of magical beliefs*. Droze.

Sandom, C., Faurby, S., Sandel, B., & Svenning, J.-C. (2014). Global late Quaternary megafauna extinctions linked to humans, not climate change. *Proceedings of the Royal Society B, 281*, 20133254. https://doi.org/10.1098/rspb.2013.3254

Satterthwaite, D. (2009). The implications of population growth and urbanisation for climate change. *Environment & Urbanisation, 21*, 545–567.

Schramski, J., Gattie, D., & Brown, J. (2015). Human domination of the biosphere: Rapid discharge of the earth-space battery foretells the future of humankind. *Proceedings of the National Academy of Sciences, 112*(31), 9511–9517.

Shane, F., Loewenstein, G., & O'Donoghue, T. (2002). Time discounting and time preference: A critical review. *Journal of Economic Literature, 40*(2), 351–401.

Sheldon, K. M., & McGregor, H. A. (2000). Extrinsic value orientation and "the tragedy of the commons". *Journal of Personality, 68*, 383–411.

Shove, E. (2003). *Comfort, cleanliness and convenience. The social organisation of normality*. Berg.

Shove, E. (2009). Everyday practice and the production and consumption of time. In E. Shove, F. Trentmann, & R. Wilk (Eds.), *Time, consumption and everyday life*. Berg.

Sidanius, J., & Pratto, F. (1999). *Social dominance: An intergroup theory of social hierarchy and oppression*. Cambridge University Press.

Slovic, A., & Tversky, A. (1974). Who accepts Savages's axiom? *Behavioural Science, 19*(6), 368–373.

Smil, V. (2011). Harvesting the biosphere: The human impact. *Population and Development Review, 37*(4), 613–636.

Smith, S., & Razzell, P. (1975). *The pools' winners*. Calibon.

Smith, E. A., & Wishnie, M. (2000). Conservation and subsistence in small-scale societies. *Annual Review of Anthropology, 29*, 493–524.

Sober, E., & Wilson, D. S. (1998). *Unto others: The evolution and psychology of unselfish behaviour*. Harvard University Press.

Solnick, S. J., & Hemenway, D. (1998). Is more always better? A survey on positional concerns. *Journal of Economic Behaviour & Organisation, 37*, 373–383.

Sorman, A. H., & Giampietro, M. (2013). The energetic metabolism of societies and the degrowth paradigm: Analysing biophysical constraints and realities. *Journal of Cleaner Production, 38*, 80–93.

Sörqvist, P., & Langeborg, L. (2019). Why people harm the environment although they try to treat it well: An evolutionary-cognitive perspective on climate compensation. *Frontiers in Psychology, 10*, 434719. https://doi.org/10.3389/fpsyg.2019.00348

Sorrell, S., Dimitropoulos, J., & Sommerville, M. (2009). Empirical estimates of the direct rebound effect: A review. *Energy Policy, 37*, 1356–1371.

Stefan, J. (2015). Influence of indoor and outdoor plants on health: Synthesis of research. *Canadian Psychology, 56*(4), 405–425.

Steffen, W., et al. (2015). Planetary boundaries: Guiding human development on a changing planet. *Science, 347*(6223), 1259855.

Steffen, W., et al. (2018). Trajectories of the earth system in the anthropocene. *PNAS, 115*(33), 8252–8259.

Stiglitz, J., Sen, A., & Fitoussi, J.-P. (2009). Report of the Commission on the Measurement of Economic Performance and Social Progress.

Summerhayes, C. P., & Zalasiewicz, J. (2018). Global warming and the anthropocene. *Geology Today, 34*(5), 194–200.

Tainter, J. A. (1988). *The collapse of complex societies*. Cambridge University Press.

Tajfel, H., Billig, M. G., Bundy, R. P., & Flament, C. (1971). Social categorisation and intergroup behaviour. *European Journal of Social Psychology, 1*, 149–178.

Taylon, B. (2010). *Dark green religion: Nature spirituality and the planetary future*. University of California Press.

Thaler, R. (1981). Some empirical evidence on dynamic inconsistency. *Economics Letters, 8*, 201–207.

Tomasello, M. (2009). *Why we cooperate*. MIT Press.

Trianer, T. (2010). Can renewables etc. solve the greenhouse problem? The negative case. *Energy Policy, 38*, 4107–4114.

Tsao, J. Y., & Waide, P. (2010). The world's appetite for light: Empirical data and trends spanning three centuries and six continents. *Leukos, 6*, 259–281.

Turchin, P., Currie, T. E., Turner, E. A. L., & Gavrilets, S. (2013). War, space, and the evolution of Old World complex societies. *Proceedings of the National Academy of Sciences USA, 110*, 16384–16389.

Uchida, Y., & Oishi, S. (2016). The happiness of individuals and the collective. *Japanese Psychological Research, 58*, 125–141.

Van de Velde, C. (2018). Presentation: Sociology of solitude: Concepts, challenges, perspectives. *Sociologie et sociétés, 50*(1), 5–20.

Van Veelen, M., & Hopfensitz, A. (2007). In love and war: Altruism, norm formation, and two different types of group selection. *Journal of Theoretical Biology, 249*, 667–680.

Veblen, T. (1898). Why is economics not an evolutionary science? *Quarterly Journal of Economics, 12*, 373–397.

Veblen, T. (1899). *The theory of the leisure class*. (1998 edition) (Great minds series). Prometheus Books.

Vereecken, W., Van Heddeghem, W., Colle, D., Pickavet, M., & Demeester, P. (2010). Overall ICT footprint and green communication technologies. Paper presented at the 4th international symposium on communications, control and signal processing, Limassol, Cyprus.

Warner, J., & Pleeter, S. (2001). The personal discount rate: Evidence from military downsizing programs. *American Economic Review, 91*, 33–53.

Weber, E. U. (2006). Experience-based and description-based perceptions of long-term risk. *Climatic Change, 77*, 103–120.

White, L. J. (1967). The historical roots of our ecologic crisis. *Science, 155*, 1203–1207.

Wilson, E. O. (1999). *The diversity of life*. Norton.

Wilson, E. O. (2012). *The social conquest of the earth*. Liveright. W.W. Norton.

Wilson, D. S., John, M., & Gowdy, J. (2013). Evolution as a general theoretical framework for economics and public policy. *Journal of Economic Behavior & Organization, 90*, S3–S10.

Wilson, M. L., Kahlenberg, S. M., Wells, M., & Wrangham, R. W. (2012). Ecological and social factors affect the occurrence and outcomes of intergroup encounters in chimpanzees. *Animal Behaviour, 83*, 277–291.

Wrangham, R. W. (2018). Two types of aggression in human evolution. *Proceedings of the National Academy of Sciences USA, 115*, 245–253.

Wright, R. (2004). *A short history of progress*. House of Anansi Press.

Zarzan, J. (2005). *Against civilization: Readings and reflections*. Feral House.

Zvelebil, M., & Rowley-Conwy, P. (1986). Foragers and farmers in Atlantic Europe. In M. Zvelebil (Ed.), *Hunters in transition* (pp. 67–96). Cambridge University Press.

GPSR Compliance

The European Union's (EU) General Product Safety Regulation (GPSR) is a set of rules that requires consumer products to be safe and our obligations to ensure this.

If you have any concerns about our products, you can contact us on ProductSafety@springernature.com

In case Publisher is established outside the EU, the EU authorized representative is:

Springer Nature Customer Service Center GmbH
Europaplatz 3
69115 Heidelberg, Germany

Batch number: 09013100

Printed by Printforce, the Netherlands